MW01462477

Modular Mansions

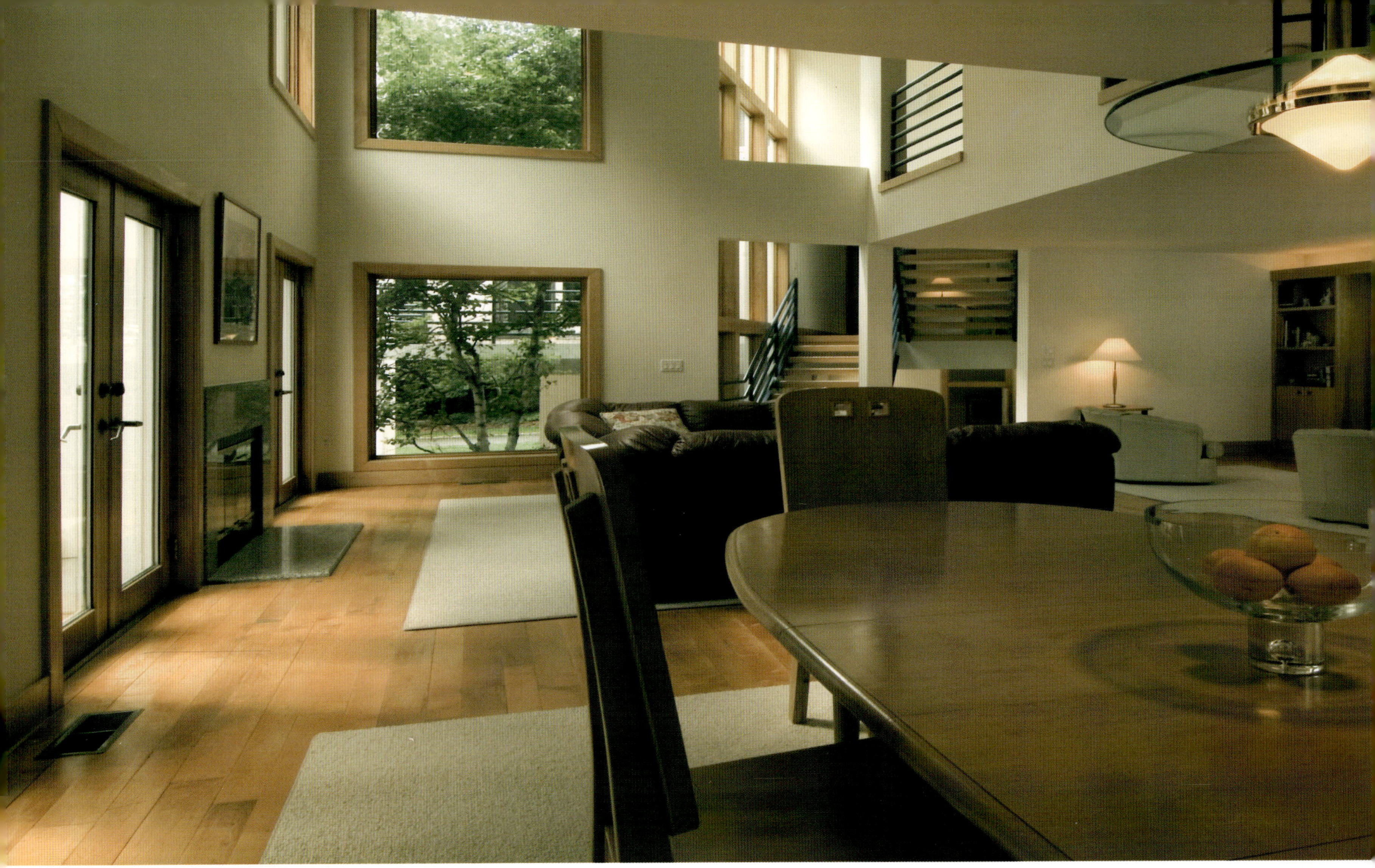

Modular Mansions

Sheri Koones

Gibbs Smith, Publisher
Salt Lake City

First Edition
09 08 07 06 05   5 4 3 2 1

Text © 2005 Sheri Koones
Photographs © 2005 Philip Jensen-Carter, except as noted on page 160

All rights reserved. No part of this book may be reproduced by any means whatsoever without written permission from the publisher, except brief portions quoted for purpose of review.

Published by
Gibbs Smith, Publisher
P.O. Box 667
Layton, Utah 84041

Orders: 1.800.748.5439
www.gibbs-smith.com

Designed by McQuiston & Boyer
Printed and bound in Hong Kong

Library of Congress Cataloging-in-Publication Data
Koones, Sheri, 1949-
  Modular mansions / Sheri Koones.— 1st ed.
      p. cm.
   ISBN 1-58685-712-6
  1. Modular construction. 2. Prefabricated houses. 3. Architecture, Domestic—United States. 4. Mansions—United States. I. Title.
   TH1098.K68 2005
    728—dc22
                        2005009956

For my mom
(who always made our house a home)
with lots of love and affection.

contents

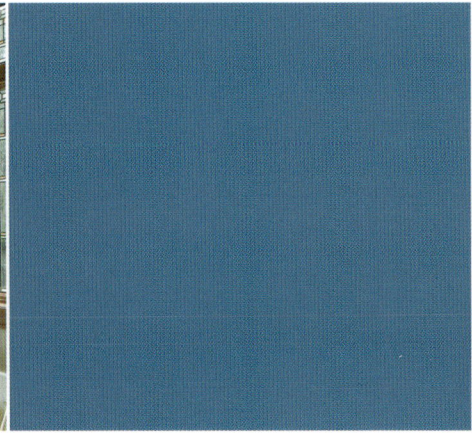

Acknowledgments 9

Introduction 11

The Construction 20

Patriot Spy Farm 30

Nantucket 36

    Federal House 40

    Slice of Life 43

Arts & Crafts House on Lake Orange 46

The Douglas Cutler House 54

The Cape House 62

Summerside Cottage 68

Lake Winnipesaukee 74

    Cottage on the Lake 76

    Governor's Island Paradise 80

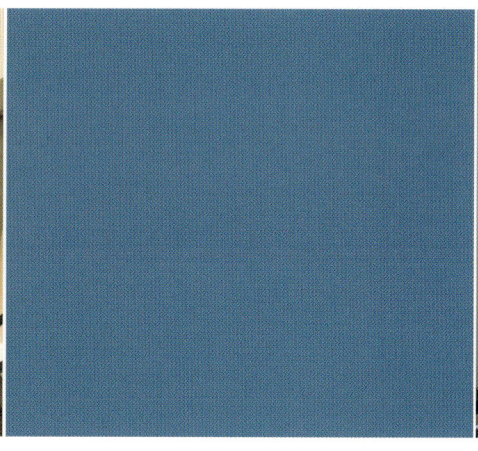

Batkins' Arts & Crafts 84

Fanelli Homestead 94

Canyon View Log Home 100

The Georgian 106

Telluride 114

    Spiritsrest 117

    Serapio Road House 120

Juniper Hill House 124

The Colonial on Lake 132

The Glide House 140

Victorian Nostalgia in New England 146

Hidden Pond Lane 152

Bibliography 158

Resources 158

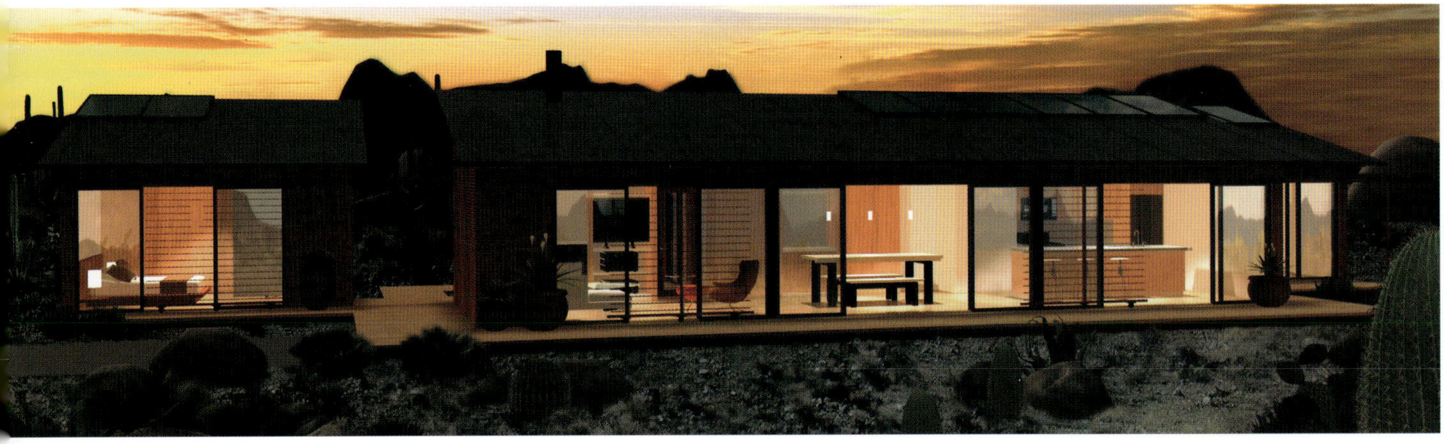

Acknowledgments

I owe a great debt to all of the homeowners who graciously opened their beautiful homes to me. Their shared experiences added so much insight and interest to this project.

Philip Jensen-Carter was a joy to work with and is the ultimate professional photographer. My thanks for a job well done to the highest standards. Thank you also to the other photographers who did a fine job.

Thank you to the builders who helped me gain access to many of the houses and graciously assisted me in getting the houses photographed. A special thank you to Curtis Barnes, the town crier, and to Charles Hensel, who showed us the best of Nantucket, and to Paul Fournier for a wonderful time in New Hampshire.

A special thanks to Dave Wrocklage, who put enormous effort into helping find several of the homes and consulted on many of the details. He made my job not only so much easier but also very pleasant.

Many thanks to the modular manufacturers who painstakingly helped me locate the projects to photograph.

My gratitude to all of the architects and interior designers who shared their great insights, experience and project details with me.

My thanks to the crew at Gibbs Smith, Publisher: Suzanne Taylor for encouraging this project, Hollie Keith for her fine editing and Jessica McKenzie for her marketing skills.

And, as always, thank you to my family and friends who are a continuing source of inspiration and encouragement.

A comfortable rocking chair on the front porch is a wonderful place to relax and read a book.

Introduction

It is surprising to find out how many people still think of modular houses as trailers. Having been exposed to beautiful modulars in my own town, I find great pleasure in showing friends some of the magnificent houses I have found in the last few years. They totally break the stereotype people have of this kind of construction.

I was first taken with the concept of modular houses when one rainy day several years ago I watched a friend's house being set. The boxes, or modules, one by one, were lifted with a crane and placed in a foundation that had been poured several weeks earlier. I returned on the second day and the house was almost completely intact. It took several months to finish the exterior and interior, but it was ready much sooner than any other house I had ever seen. Watching this event caught my interest and I have been a proponent ever since.

Modular housing is a growing trend. Whereas in the initial years of production the modules were generally simple boxes, modular construction has become significantly more sophisticated, often producing houses that are impossible to differentiate from traditional stick-constructed ones.

In 1990, there were twenty thousand modular houses built in the United States. By 2003, there were almost thirty-eight thousand modular homes built (according to an annual survey on modular construction by Hallahan Associates, an industry consulting firm). This number continues to grow each year, with a larger portion of modular homes being upscale.

As people are becoming more aware of all the advantages of modular construction, it has begun to emerge as a popular method of building custom homes along with the more modest-priced houses of the past. Beautiful, spacious homes have been constructed in the last several years that rival the size and quality of many custom stick-built homes.

Modular homes are sometimes confused with HUD code housing, also referred to as manufactured homes. This type of construction emerged from the production of mobile-type homes. Built on metal chassis with shallow roof pitches and boxy features, manufactured homes are transported on their own wheels. The emergence of these homes gave rise to more affordable homes for the masses. All HUD homes are built to meet strict government code requirements. As that business has been on the decline in recent years, many HUD construction companies have started building modular houses.

The boxes of modular construction are stick-built in a factory and transported on trucks to a site where two or more are attached to each other and set on a prepared foundation. For the most part, they are built very much like stick houses, in all the styles and with all the amenities; the only difference is that they are built in a factory. There are two types of modular houses that can be purchased today: those that are offered by the manufacturer, with some minor changes and upgrades, and the more customized homes that are individually designed. This last category has recently become a substantial part of the business. Several very creative architects and designers have become interested in designing modular homes, and the designs have become more sophisticated as a result. While years ago modular homes were built in a very limited array of styles, today they can look like almost any style and come in any size.

Modular companies differ in the services they offer. Some companies design and construct the house, prepare the foundation, and complete all of the finishing work required. Other companies design and build the house, but the homeowner must hire a builder to prepare the foundation and do all of the required site work. Some companies offer specific designs and/or design services; with others, homeowners must employ the services of an architect if they desire a more customized plan. Most companies require that the modules be purchased through a builder, although there are some companies that will sell directly to consumers. For people who aren't familiar with local builders, modular companies can often suggest builders they have worked with in the area. Builder consultants are available to recommend architects, manufacturers and builders around the United States and Canada. The selection of professionals will depend on location, time constraints and the complexity of the design.

There are several reasons why modular houses have become so popular. One reason most commonly given by homeowners is the time involved in building a modular home compared to the time involved in building a stick-built house on-site. While the modules are constructed in the factory, a crew can simultaneously prepare the foundation and install the utilities at the house site. Whereas it generally takes at least a year to complete a site-built house, a modular one can be finished in three to eight months, depending on the complexity, size and location of the project. Because of the savings in construction time, there is also a savings in the time the homeowner will have to carry a construction loan and maintain an additional residence. Often a family must move out of their current home for a transition period until their new home is complete, and there are rental costs and storage costs in addition to the inconvenience.

Snow, rain and extremely hot or cold weather often delay construction schedules for stick-built homes significantly, particularly in certain areas of the country. In the Hamptons on Long Island, for example, there can be periods of blowing sand, making it impossible to construct a house.

The cedar throughout the Canyon View Log Home (see page 100) adds warmth and creates the feeling of a traditional log home.

Because modular houses are constructed in a protected environment, there will be minimal weather delays, allowing the modular company to stay on schedule.

Being constructed in a climate-controlled environment, the building materials are protected from exposure to the elements. Site-built homes can take months to close up, often compromising the materials with all of the wear and tear caused by exposure to wind, rain, snow and extremes in temperature. Modular homes can usually be made weather tight within a day or two of being set.

Wood used by modular companies generally has time to acclimate to the environment before it is used. The wood usually comes from one lot, which means all the wood in the house will weather at the same rate, limiting the number of cracks that can sometimes be found in stick-built houses. Most modular companies use kiln-dried wood instead of green wood, which also significantly reduces the shrinkage that will occur later.

The factories that produce modules have become more technologically advanced with state-of-the-art equipment that would be impractical to use with on-site construction. Saws that make multiple cuts and other state-of-the-art machinery are commonly found in the newest plants. Some companies have also begun to incorporate the latest advances in safety and energy efficiency.

Modular components travel at high speeds on flatbed trucks to reach their destination. When they are set, they are lifted by a crane into place. Because of the stress put on them, the modules must be built particularly strong to endure these rigors. Builders have told me that their on-site homes could never survive the rough conditions that modular homes must endure. Modular sections generally contain more building materials than comparable site-built homes, sometimes with multiple methods of attaching drywall and triple headers used over window and door openings. The walls between modules are doubled where two boxes are attached. This is true for the floors as well, significantly adding to the soundproofing and strength of the house.

The cost of a modular house is usually between 10 and 15 percent less than a similar site-built house. Companies are able to charge less because they use local craftsmen and contractors whose wages are less than those paid in some of the most affluent areas of the country. Craftsmen are kept busy twelve months a year, moving from project to project without the waste of time that often occurs when a builder is constructing an individual home on-site. Materials are usually bought in greater quantities, allowing the modular company to pass those savings to the consumer. There is also less waste of materials when they are cut in a plant and less pilferage than can occur on open lots. Modular companies generally prefer to use particular brands of certain items, such as windows, because they are able to buy them in large quantities and can get better prices. For a higher charge, however, many companies will offer a variety of brands.

Because construction time is reduced, the homeowner pays interest on a construction loan for a shorter period. Whereas a site-built home may take twelve to fourteen months to build, a modular house will generally be complete in three to eight months at most. This can mean six months or more of savings in interest. Some banks will inspect the home after it has been set to make sure it has been done correctly. This gives the homeowner an added protection against future problems. Not all loan companies are familiar enough with modular construction to adequately offer this service, but it is wise to find a loan officer who is experienced in this area, if possible. When an additional rent or mortgage must be paid during construction, that cost is also reduced by the shortened time it takes to build a modular home.

At one time the biggest deterrent to building modular was the limitations in design options. Companies offered limited designs to keep their costs down, and for the most part houses were purchased by consumers looking for an inexpensive alternative to on-site construction. In the last several years, consumers have started to see the advantages to modular construction beyond just the cost savings, resulting in more companies being willing to build the grander, more interesting customized styles. Today, several companies are building those houses that are out of the box—literally. *Building Systems* magazine recently featured a thirty-three-module house. Although not quite the norm, larger houses are becoming more common in several areas of the country. All of the design features and amenities that have been available on stick-built houses are now available on modular ones—hot tubs, fireplaces, vaulted ceilings, multiple rooflines, custom kitchens, upscale lighting and so on.

The role of the architect in modular design is also changing with the emergence of larger, custom-designed houses. For many years modular companies have offered a group of plans from which the homeowner could select. Often they would tweak the design to suit the individual customer's needs. As the houses have become grander and more sophisticated, homeowners have begun to look for architects qualified to design their modular houses. Currently, more architects are becoming knowledgeable of the design requirements of modular housing (i.e., with the limitation in box size and the need for proper orientation of the boxes) and have easily adapted to this process to meet the demand for more professional designs.

The documentation requirements are somewhat different with modular homes. Architects do not have to submit working drawings because that is done by the modular company. Architects are still required to assist in securing special permits (other than the basic permits), do a shop review of the completed documents, and execute designs for interior trim and lighting. The review of shop drawing is even more critical with modular homes because of the difficulty of changing designs after the final approval is given; site-built homes are easier to alter during construction. Many companies today employ their own architects to work with customers in altering standard designs or developing custom designs. Homeowners can opt to select a company with an architect on staff, or they can employ their own.

The warranty offered on modular homes is another incentive to build with this method. Rarely will a site-built house come with longer than a one-year warranty, whereas modular companies generally offer a ten-year warranty on their homes. Most modular companies utilize maintenance companies around the country that are familiar with their homes and will fix problems that may occur.

Inspections are also done at the plant by independent third-party engineering companies that ensure code compliance in the municipality where the house will be set. Modules are inspected once or twice a week at the factory, avoiding the delays that sometimes occur on-site when builders may have to wait days or weeks for a local inspection.

Several companies around the country have become "vertical" operations; they cut down their own trees, mill their own wood and construct the house as well. There is a new company forming on the Hoopa Reservation in northern California that will be run by the Hoopa Tribe. The area abounds with large, beautiful redwood trees that will be used for the houses they will build. The West Coast of the United States has lagged behind the rest of the country in offering modular construction. Only now are companies seeing the potential on the West Coast and building plants to meet growing demands.

There are some lots with difficult access, which makes setting modules more challenging. There are homes, however, that have been set on cliffs and very steep lots. One manufacturer described a road where the widened load could not pass and a helicopter was employed to set the boxes in place.

Most modular companies will ship only to a designated area. For some companies, it is a certain group of states; for others, it is a defined distance. Most companies will not ship beyond 1,600 miles; other companies have an even shorter range. The farther the distance the modules must travel, the greater the expense. Some areas of the country are so far from any supplier that it is not practical to pay this added cost. Companies must find large and secure locations to store the units until they are ready to be set. Because the units are so large and can potentially clog roadways, some municipalities have adapted a lottery system to limit the amount of larger boxes coming into a particular area. Permits are required and may cause a delay in the schedule. This can put a strain on the modular company and delay the move-in time for the homeowner.

Modular construction is still in its early stages of development when compared to its incredible potential. With the dwindling population of craftsmen, the sophistication of available equipment, and the added desire by so many for faster construction solutions, modular construction is apt to grow at an even faster rate in the future. New regulations will be necessary, however, to establish consistent code regulations in all states, making it easier for modular companies to deliver houses across state lines.

Fans on the rear overhangs of the Fanelli Homestead (see page 94) move the warm air in the summer to make it more comfortable to sit outside under the gazebo.

The Construction

WATCHING A HOUSE being built in a factory is a unique experience. The factories that build modular houses are immense, with many more employees than would ever be seen around a typical construction site. All employees wear protective eyewear required by law. Factories are very busy places with several people working in various areas. The boxes, or modules, move from area to area as sections of the box are completed. In some plants they move once a day, with workers rotating from section to section, while at other plants the sections move regularly throughout the day.

There are generally many houses in production at the same time. Carpenters, plumbers and electricians are always present and are available to work on each house as it is ready. There is no waiting for a craftsman to show up in order to complete another part of the house. Craftspeople move from box to box, completing their portion of the work. There can be a dozen people

Opposite page Workers are preparing a module for the installation of the ceiling system.

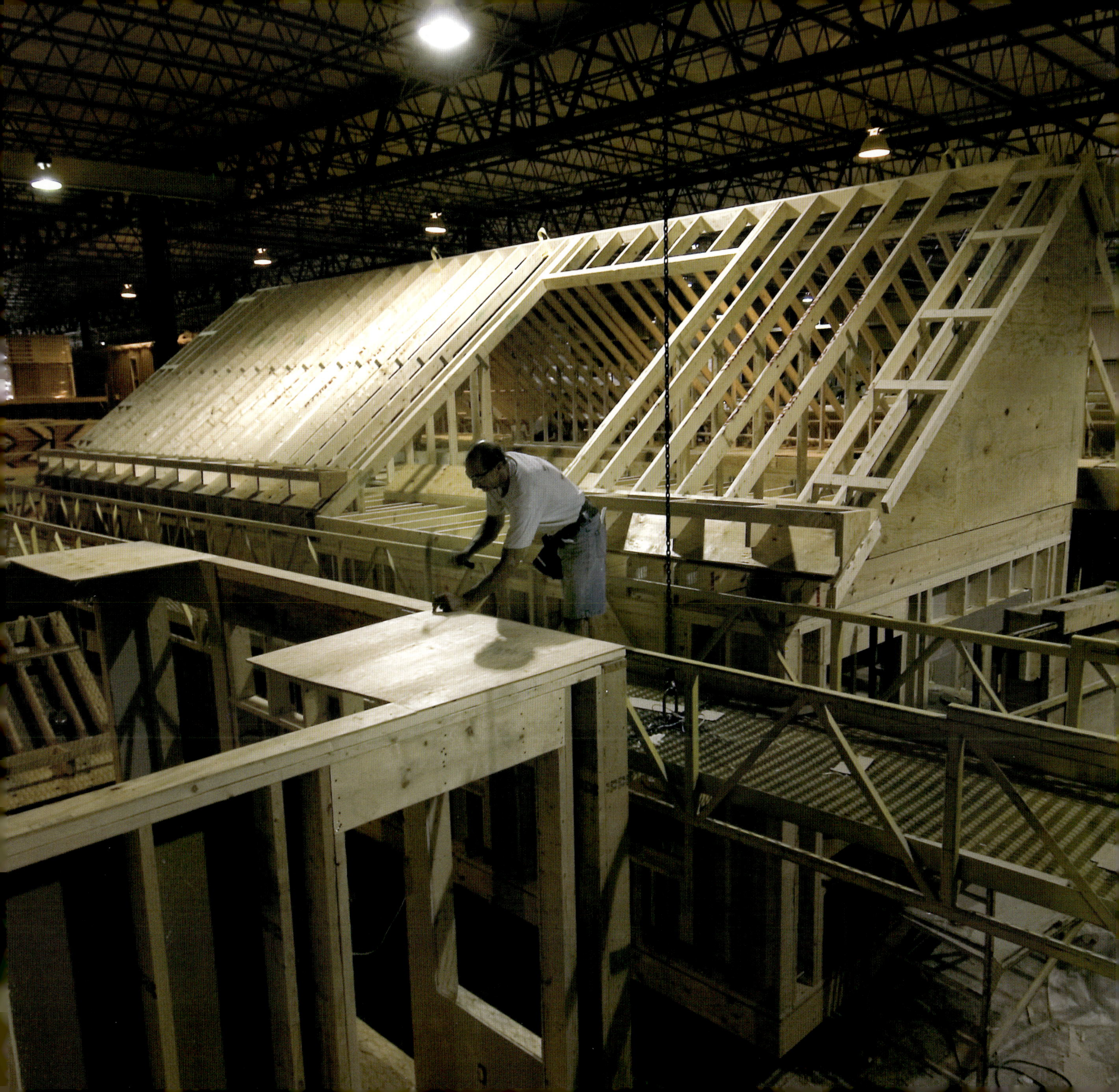

This roof system will be set on a two-story house or a ranch-style house.

working on a box at any given time. Homeowners who have visited plants have expressed excitement over seeing so many people working on their home at one time.

Typically, factories can produce five to seven houses a week, depending on the company. It can take six to ten days for a house to be completed from the time the lumber is cut until the boxes are out the door. Some intricate homes may take a bit longer, but rarely will take more than two weeks. That time schedule is a miracle to anyone who has watched their home slowly being built on-site. Generally there is a one- to two-week delay before the boxes can be delivered; longer delays can occur for areas where arrangements with the municipality may be more difficult.

The process begins in a section of the factory where the wood is precut. The worker is given a "cut list" and saws a good portion of the wood needed for that house. Some factories use the typical saws seen with on-site work, while more sophisticated factories now employ saws that can be

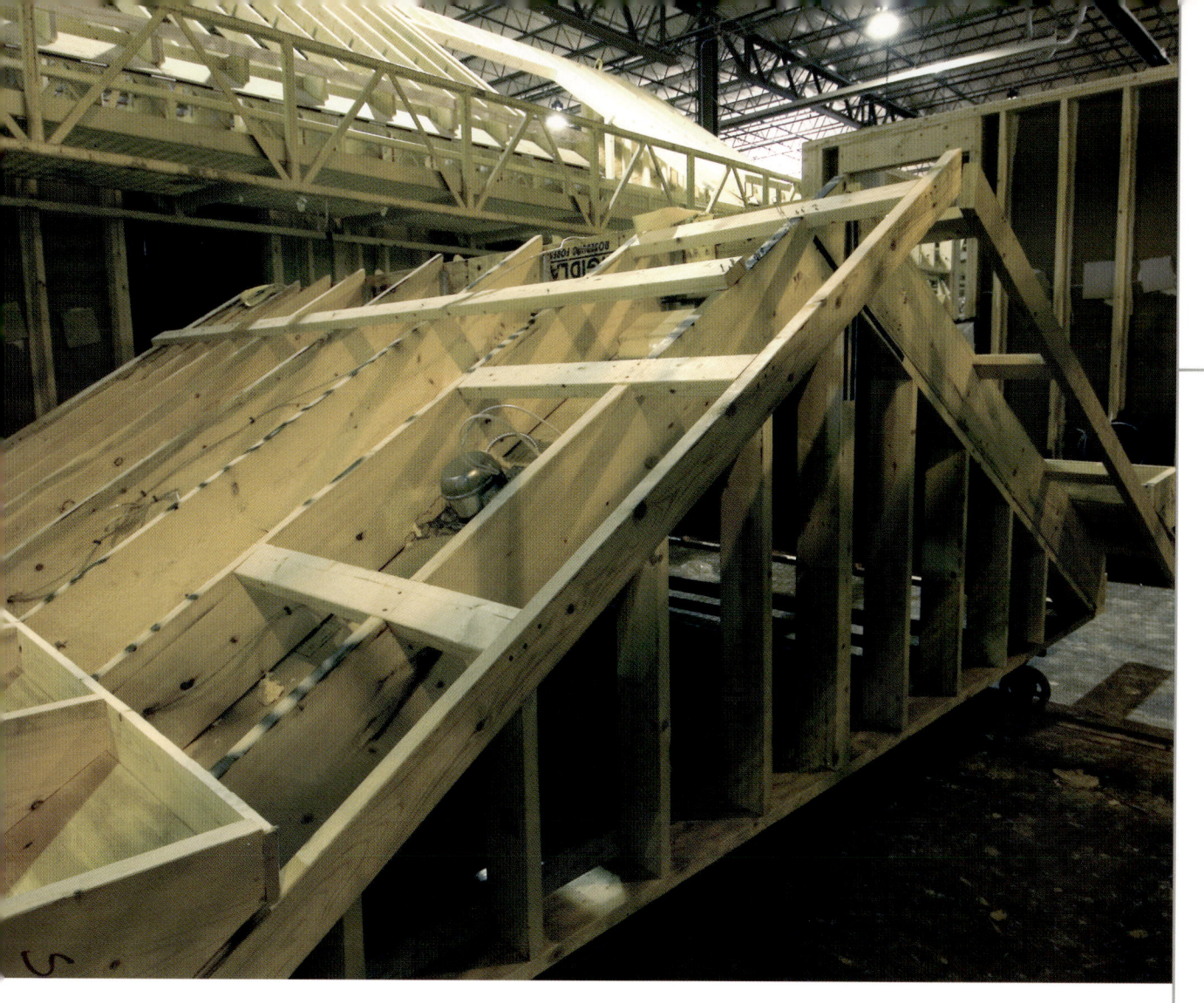

Roof "bonnets" are individual roof sections that are attached separately when the home is set.

programmed and are able to make up to five cuts in the wood at the same time. Kiln-dried wood is used instead of green wood, which must be air-dried over several weeks or months, depending on local weather conditions. Because kiln-dried wood is used, the frame of the house is not subjected to twisting and shrinking.

The floors are framed first on a platform. Clips are used to make sure the wood is very square. The size of the lumber will vary depending on the load it will have to carry. A room that will have a granite countertop, for example, will have to support a heavier load and will require a heavier frame. Lumber can vary from two-by-fours to two-by-tens. Sometimes a double or triple perimeter of wood is used for added strength.

The next part of the house to be built is the walls, which in some factories are built upright and in

"watching a house being built in a factory is a unique experience"

others built flat. In those factories where they are built flat, the studs are laid out on a large steel jig to keep them straight and square. When the walls are complete, they are lifted in sections with an overhead crane and quickly attached to the floor section. Unlike houses built on-site, the Sheetrock for the interior walls is generally attached to the studs before the sheathing. On-site, the exterior walls are installed first in order to close up the house and make it weather tight. Factories find it advantageous to install the interior walls first because it is easier to work from the outside of the boxes, installing pipes, wires and insulation. In addition, by installing the interior walls first, the back seams of the Sheetrock can be taped, creating a stronger seal, and in some cases, the insulation is glued to the wall, avoiding the slippage that may otherwise occur. Some factories today use a foam sealant to attach the drywall or Sheetrock to the studs because it is strong and requires the use of fewer nails, which can potentially interfere with interior wires or plumbing. Traditionally this sealant has only been used for attaching ceilings. The means of attaching drywall vary from company to company and can include nails, screws, foam or any combination of these. Fire walls, which are required by code in all municipalities, cannot have the foam sealant. Residential building codes require a one-hour separation between the garage and the main house, including walls and doors, and there is a potential that the foam sealant could melt in a fire.

Each box has a checklist that is controlled by one or more supervisors in the plant. They check each step in the production and sign off on procedures that are satisfactorily completed. This way there is tight control on the work being done by a large staff.

The ceiling sections are built on a horizontal surface, with the joist framework laid out over the Sheetrock and then attached with a foam spray that creates an instantaneous and strong bond. The ceiling sections are also put into place with the use of overhead cranes.

After the Sheetrock is attached, the windows and doors are routed out. Electricians run wires in the open walls. It is interesting to see workers on stilts joint-compounding the seams. This compound is often referred to as Spackle, or mudding, the slang often used by the workers. Propane heaters and fans are used to speed up the drying process, which may take several days.

Roofs are built with hinges so they can be shipped flat and lifted to the correct slope on-site. When the roof section is complete, it is lifted by a crane, placed on top of a module and nailed securely in place. Roofing materials are installed in the factory for some houses; for others, they are installed by the builder after the house has been set. Asphalt is the most commonly used roofing material

The insulation is installed from the exterior of the house before the siding is placed.

When code requires an air infiltration barrier, it is applied at the factory.

on modular houses, but other materials are also currently being installed on custom homes, such as metal, wood shingles and shakes.

Certain items are more difficult for modular companies to produce. Curved walls, for example, are difficult to build in the factory, and only a few companies that specialize in custom homes are willing to include this feature. Whenever the factories have to produce an out-of-the-ordinary feature, it will hold up the production line and cause delays.

Next, the sheathing is installed. The walls are taped and then mud-ded or compounded. The interior finish work is done and kitchens are installed if they are to be completed at the factory. Most of the flooring is installed on-site with the exception of some wood floors. Most companies do not employ masons to install tile either, so this installation is often done on-site as well. The boxes are then given a primer coat of paint. Generally, the finish coat is done on-site.

Windows and some siding are installed at the factory. Wood clapboard, composite and vinyl siding are often installed in the factory, while other materials such as wood shingles, brick or stucco must be installed on-site by the builder.

There are several items not included in the factory production. Metal inset or gas fireplaces can be installed in the factory, but masonry fireplaces are too heavy and must be installed on-site. Some homeowners choose to receive their home with kitchen cabinets installed; others prefer their own custom cabinets to be completed on-site. Sometimes dish-

Modules are covered with tarps and put in the yard, awaiting delivery to the construction site, where they will be set.

washers are shipped with the units because they are labor intensive to install, but often all appliances are installed on-site.

Although wiring and vents are installed for the mechanical systems, none of the furnaces, boilers, air conditioners or water heaters are installed in the factory. The wiring for audio, lighting and security systems can be installed at the factory, but the keypads, fixtures and speakers, along with additional equipment, are installed on-site. Central vacuum-system piping can be installed but must be laid out carefully in the planning stage. Planning, in general, must be done more carefully for modular homes because of the difficulty in making changes after the production has begun.

There are several engineering companies around the country who will inspect modular houses at the factory. They are paid by the modular company and generally come once or twice a week. Like local inspectors, they check all the wiring and plumbing and make sure that the house meets the codes required by municipalities around the country. Because the houses are inspected during all phases of the production, the walls can be delivered closed. Local inspections are later done on-site.

After the modules have been swept clean, they are wrapped in plastic and prepared for the trip to the site where they will be set. Each module is loaded on a flatbed truck for shipment to the location, which may be nearby or several states away. For some homes, such as several in this book, the boxes must be transferred to barges to be sent to an island and then set.

Ninety percent of a modular house can be completed in the factory, although with many custom homes, owners opt to complete a larger percentage on-site.

Modules vary in size, with the largest that can be transported in most areas being sixteen feet wide and sixty feet long. Modules can be built eight feet high if the roof section will be included. If the sections will be shipped without roofs, the sections can be nine feet high. Some states, such as Texas, allow the boxes to be twenty by ninety feet. States vary in their regulations for transporting modular houses. In New York, for example, modules must be out of the city before four o'clock in the afternoon. Connecticut will only allow sixteen of the widest units to pass through the state per week, and the state uses a lottery system with which all modular companies must comply.

The true miracle of modular homes is how they are set. A crane lifts the individual sections, and a

This completed module is ready to be delivered and set.

crew then attaches each section to the others. In a day or two, most houses can be weather tight. With very large houses, this process may take slightly longer. The sections are bolted together, and the roof sections are lifted onto the house.

For neighbors who have been away for a few days or have not passed a particular street regularly, they are shocked to see a house where none was standing just a few days earlier. Those lucky enough to view the setting event are often "wowed" by the sight of the modules being lifted in the air and set in place, like a jigsaw puzzle.

Patriot Spy Farm

NOT ONLY DID the Macierowskis purchase 1,200 feet of frontage on the Connecticut River, but they also acquired a historical landmark at the same time. Marked by a monument erected by the Daughters of the American Revolution (DAR), this property is forever a testament to Daniel Bissell, an American patriot who spied for the country and received the Military Badge of Merit. It is the birthplace of Bissell (born in 1754) and also the original site of the historic Bissell Ferry, established in 1641, which ran from Windsor to the Boston area. The ferry was originally established by the elder John Bissell in order to graze his cattle on the other side of the Connecticut River; it later became an economic and transportation link between Boston on the east side and Hartford, New Haven and New York on the west side. It is further noted that George Washington wrote in his diary on October 21, 1789, that he visited with Oliver Ellsworth, the first chief justice of the Supreme Court, whose

Opposite page The module is lowered onto the foundation with a crane.

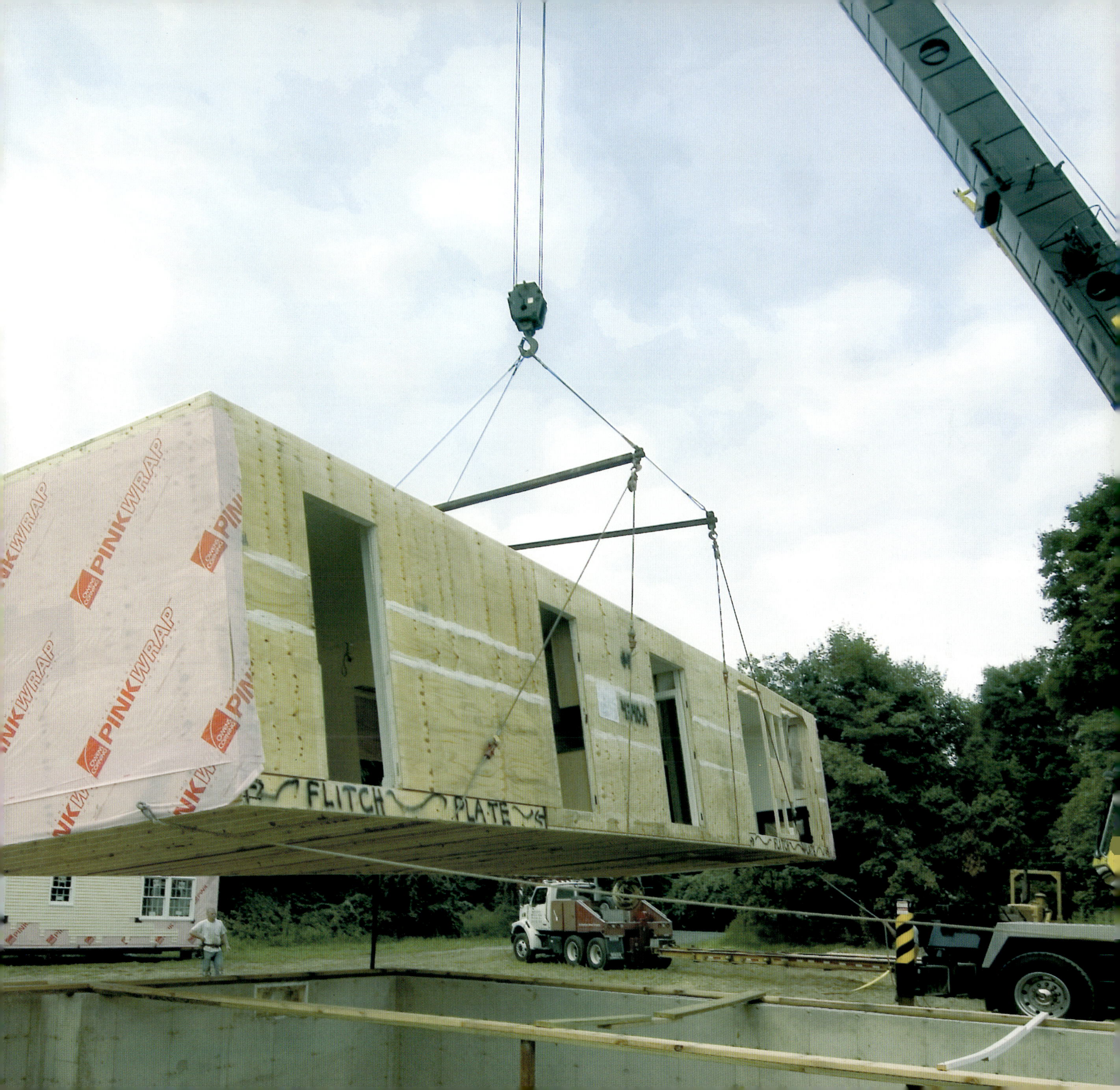
PINKWRAP
FLITCH PLATE

Above left The extension of the perimeter beam (in the corner) maintains the structural integrity of the house until it is set into place, at which time it is cut away.
Above right With the lower level set, the second floor is lifted into place.
Right A separate roof section, called a "bonnet," is set in place.

Architect:	Michael MacDonald
Manufacturer:	Westchester Modular Homes
Square feet:	4,000
Number of boxes:	7
Location:	Windsor, CT

house is still standing just a short way down the road from the Macierowski's home.

Ted and Jen were looking for property where they could live and have a small horse farm. Jen, having grown up on a farm and having always loved riding, wanted an area where she could cultivate that interest. She and Ted found thirty acres in Windsor, whose historic district is just south of their property; it was

The crew began setting the modules in the morning and by midday the house appears almost complete.

"we wanted a traditional house, in keeping with the tradition of the area"

Above The monument erected by the Daughters of the American Revolution marks the location where patriot spy David Bissell was born.

Right Several days after setting, the house looks almost complete, but the balance of the interior work and landscaping remain to be finished and siding must be installed on-site to the exterior of the garage.

the first settled town in Connecticut.

After much investigation, the Macierowskis chose to build a modular home because they wanted to avoid months of delays in construction and cost overruns typical with on-site construction. With the help of their local architect, Michael MacDonald, they altered the manufacturer's plans to fit their personal needs and to take the best possible advantage of river views. MacDonald also helped to place the house on the site to achieve the best orientation to the sun and to distance the house from the road.

The house took six days to build in the factory and two days to set. Because of the size of the property, the modular company was able to store the boxes right at the site where they would be set. In a quick and

The house is in the final stages of completion several months after being set.

efficient manner, most of the boxes were placed on the first day, with the garage set the second. After approximately two months, Ted and Jen were ready to move in. They plan to build a riding ring and a barn to accommodate the horses they will keep on the property. They own a small utility vehicle they use to scout around the property and to drive to the river to look at the view or do some fishing.

Jen says she chose the center hall Colonial style "because we're in a historic district, and we wanted a traditional house, in keeping with the tradition of the area."

The Patriot Spy Farm house is a perfect example of modern technology meeting early American history.

Nantucket

A SMALL ISLAND, fourteen-and-a-half miles long by three to five miles wide, Nantucket has some of the most valuable real estate in the United States—it is no wonder, with its beautiful ocean views, historic cobblestone streets, majestic lighthouses and charming architecture. The hydrangeas here grow larger and more beautiful than almost anywhere else, and there is a peace that pervades when one arrives by ferry, private boat or plane onto this enchanting island. Many writers have been inspired by the island, including Sinclair Lewis, Truman Capote and John Steinbeck. This is a very relaxed location; there are no traffic lights, fast-food restaurants or neon signs. History abounds in Nantucket and can be seen by visiting the oldest house built there in 1686, now an exhibit open to the public, or by going to some of the other historic museums. There are some surprising tidbits about the island—both Benjamin Franklin's mother and the future owner of the R. H. Macy department store were born there. Forty-six percent of the land is protected and

Opposite page Boating is a popular pastime on Nantucket.

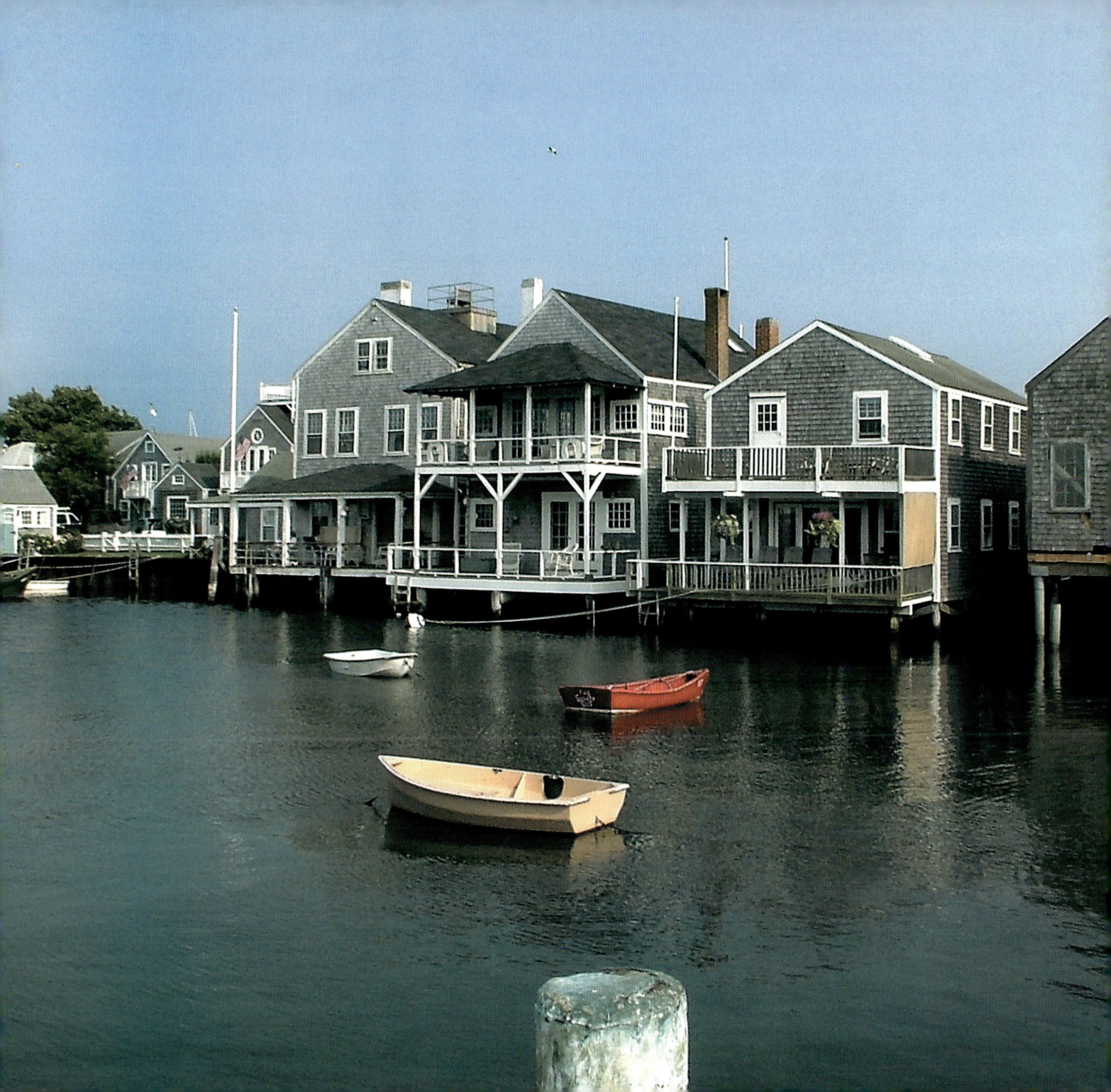

This Federal-style house is an excellent example of the architectural style and is indistinguishable from the historic houses on the island.

will never be developed, maintaining an open feeling to the island. Bicycle riding is the preferred mode of transportation, and Nantucket has charming stores and some of the best chowder one can find anywhere. What started out primarily as a whaling center has become a popular resort destination.

The local government has established a strict set of guidelines for construction in Nantucket in order to preserve the historic character of the town. These are carefully described in the book *Building with Nantucket in Mind: Guidelines for Protecting the Historic Architecture and Landscape of Nantucket Island* by J. Christopher Lang and Kate Stout. Nantucket has one of the largest collections of historic buildings (2,400) in the United States. Because of the strict construction guidelines set by the local government, it is sometimes difficult to distinguish the old from the new on the island. Here, the new harmonizes with the old.

Two fine examples of traditional Nantucket architecture are the following modular houses, built in New

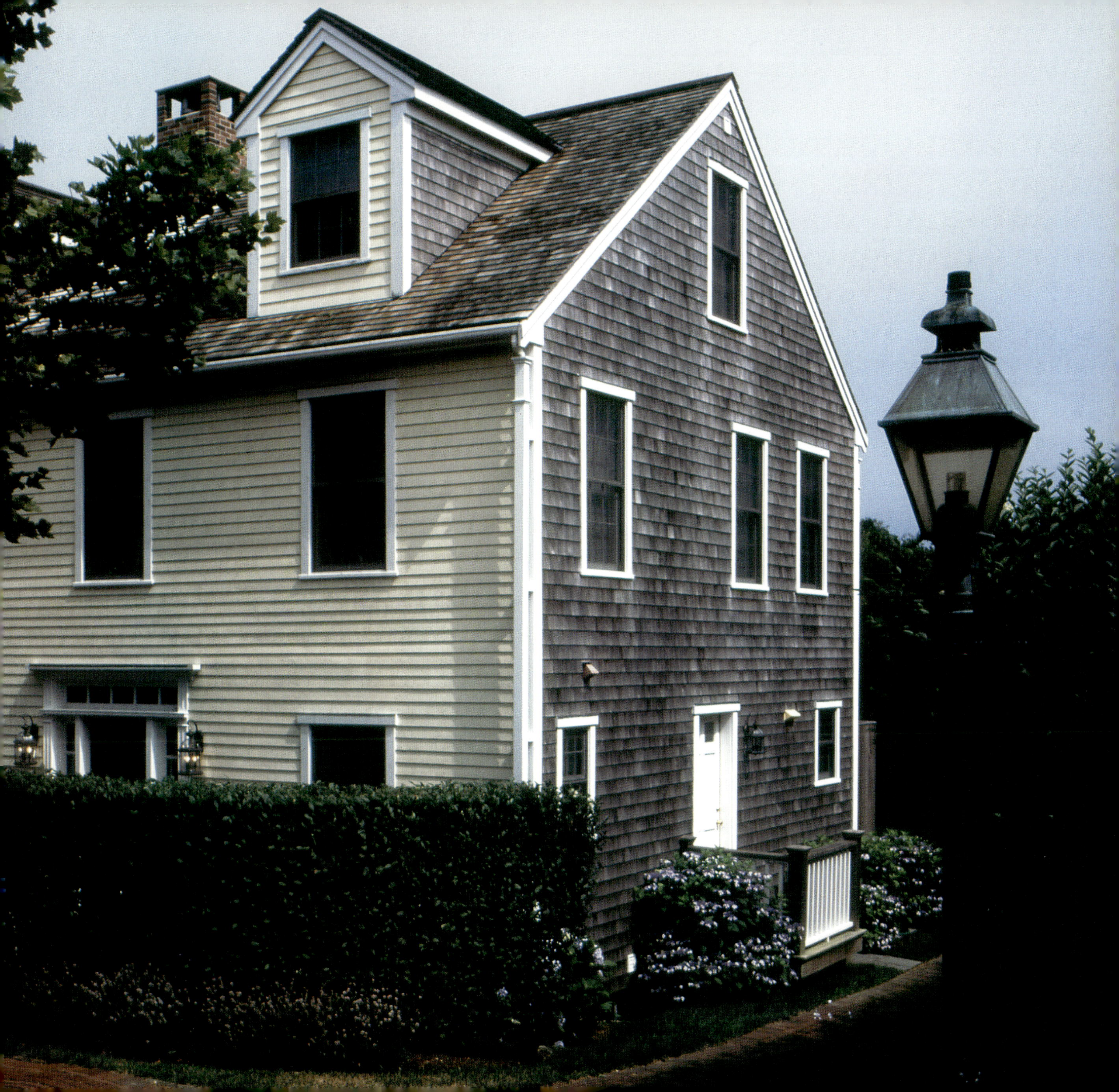

The front and rear dormers were shipped on a barge with the boxes and installed on-site.

Hampshire and sent over on a barge to the island. Some homes have required more than one trip, since the barges can only accommodate about four boxes. The modules are loaded on the barge in the evening and unloaded in the early morning hours before the other boat traffic begins. It is sometimes difficult to navigate the narrow streets with these oversized loads, but in the last several years many modules have been successfully transported and set on the island. The popularity of this type of construction in Nantucket is due to the difficulty of obtaining materials locally and finding qualified professionals, and homeowners' desire for a quick and more economical construction solution.

Federal House

> **Manufacturer:** *Epoch Homes*
> **Builder:** *Nantucket Island Homes*
> **Square feet:** *2,400*
> **Number of boxes:** *6, plus 4 dormers*

This beautiful Federal-style home is one of the most popular styles on the island. The house easily fits in with the many historic houses with gabled roofs, central chimneys and simple facades. The painted clapboard is one of the materials approved by the town to be "appropriate to the period and style"* of Federal houses. It covers the front of the house in a pale yellow; the sides are more natural windswept-gray shingles. The four dormers add much charm and architectural interest to the simplicity of the style. Since the guidelines discourage the use of shutters, the windows are kept simple with six-over-six panes on traditional double-hung windows. Built in 2000, it is used as a vacation home and rented to vacationers when the owner is not using it. The decor is traditional, consistent with the exterior design, and purposely comfortable to meet the demand of continuous vacationers over the warmer months. Like many of the houses in Nantucket, the landscaping includes beautiful pink and blue hydrangeas, which are often snipped and brought into the house to be displayed, both fresh and dried.

Building with Nantucket in Mind by J. Christopher Lang and Kate Stout.

The dining area is bright with light sweeping in on two sides.

High hedges give the house needed privacy, as it is set close to the road like most houses in town.

Passersby would never dream that this house, which looks like it has been there since the whaling days, was actually modular built.

SLICE OF LIFE

Manufacturer: *Epoch Homes*

Square feet: *2,900*

Number of boxes: *7*

A traditional design on Nantucket, the Federal style of this house fits well with the local architecture. A particularly significant feature is the roof walk, which is by definition a railed observation platform atop a usually coastal house. *Roof walk* is the politically correct term for what was once called a *widow's walk*. Since roof walks are traditionally found only on houses that are two or more stories, they cannot be built on houses that are one or one-and-a-half stories high. There

Left This comfortable living room has marine ornaments, a delightful reminder of the nautical nature of the island.

Above, top The architecture is typical of those on the island and is indistinguishable from the surrounding site-built homes. The roof walk is traditionally seen on many Nantucket homes.

Above, bottom The checkered-tile entranceway hints at the cheerful decor throughout the house.

Right This cozy bedroom opens to a small deck for sitting and reading on warm afternoons.
Below The back patio extends the living space, with the high hedges adding a degree of privacy.

are many tales associated with why these structures were built, a favorite being that they were used by women who wanted to watch for their sailor husbands returning home from sea. On the roof walks they could see ships about a day away. Today roof walks can be seen on many of the homes in Nantucket and are a charming reminder of the great history associated with the island. Like many other houses on the island, this house has been named by the owner. It is called Slice of Life, and it has a quarterboard displaying the name above the door. These quarterboards are affixed to many houses in Nantucket and come from the seafaring tradition of the island. Quarterboards are the plaques used on boats to display their names. *Slice of Life*, the name of the house, comes from the owner's business of producing those wonderful slices of old-fashioned fruit jelly candy that come in half rounds.

The traditional and antique furnishings go beautifully with the style of Nantucket design. A black-and-white tile entranceway only hints at the upbeat and lively design of the common areas of the house. It is cheerful, with ice cream colors and playful embellishments, such as the tiny chairs on the wall and the graduating-size baskets.

When the house was set in 1994, cathedral ceilings, such as the one in the great room, were unusual for Nantucket. The owner, Nancy Kissam, wanted a room with a great deal of light coming in because it would most often be used in the warmer months. Many of the houses she had looked at had very small rooms and were dark. Kissam opted for an open floor plan so that she could still interact with her family and guests when she was cooking in the kitchen.

The large wraparound deck was designed with several different areas so guests could use the spaces for sunning and quiet time, while others could be barbecuing and conversing.

The many building restrictions on the island also dictate the few acceptable coverings for driveways: pavers, stones or crushed clam shells. This house has the crushed shells, which adds to its special Nantucket character and charm.

Left The owners' whimsical collections are displayed around the house, including miniature chairs, baskets and dishes.
Right The double-high ceiling gives a great feeling of openness.

Arts & Crafts House on Lake Orange

DRIVING DOWN THIS quiet country road in Newburgh, New York, one would never expect to find this beautiful, expansive home on Lake Orange. Many old and run-down cottages appear along the road before this charming retreat becomes visible.

Owner Bob Steele grew up in Newburgh and, although he moved to Bucks County after college, his mother and brother remained in Newburgh at their homes on the lake. Many of Bob's friends left the area but later returned and either bought their family houses or bought new houses and settled in the area. Bob watched an old friend of his rebuild a Victorian in town and it gave him the impetus to want to do this as well. When a five-hundred-foot parcel of lakefront property became available several years ago, the Steeles grabbed it. Bob approached his longtime friend Scott Webb, a broker for Haven Homes, about building a vacation house on the land. Scott, who has built almost four hundred modular houses in his years in business,

Opposite page The beautiful small porch with trim, hand-crafted light fixture and Frank Lloyd Wright glass collection (Andersen Windows & Doors) inset in the door and sidelights is a wonderful introduction to this charming Arts and Crafts house.

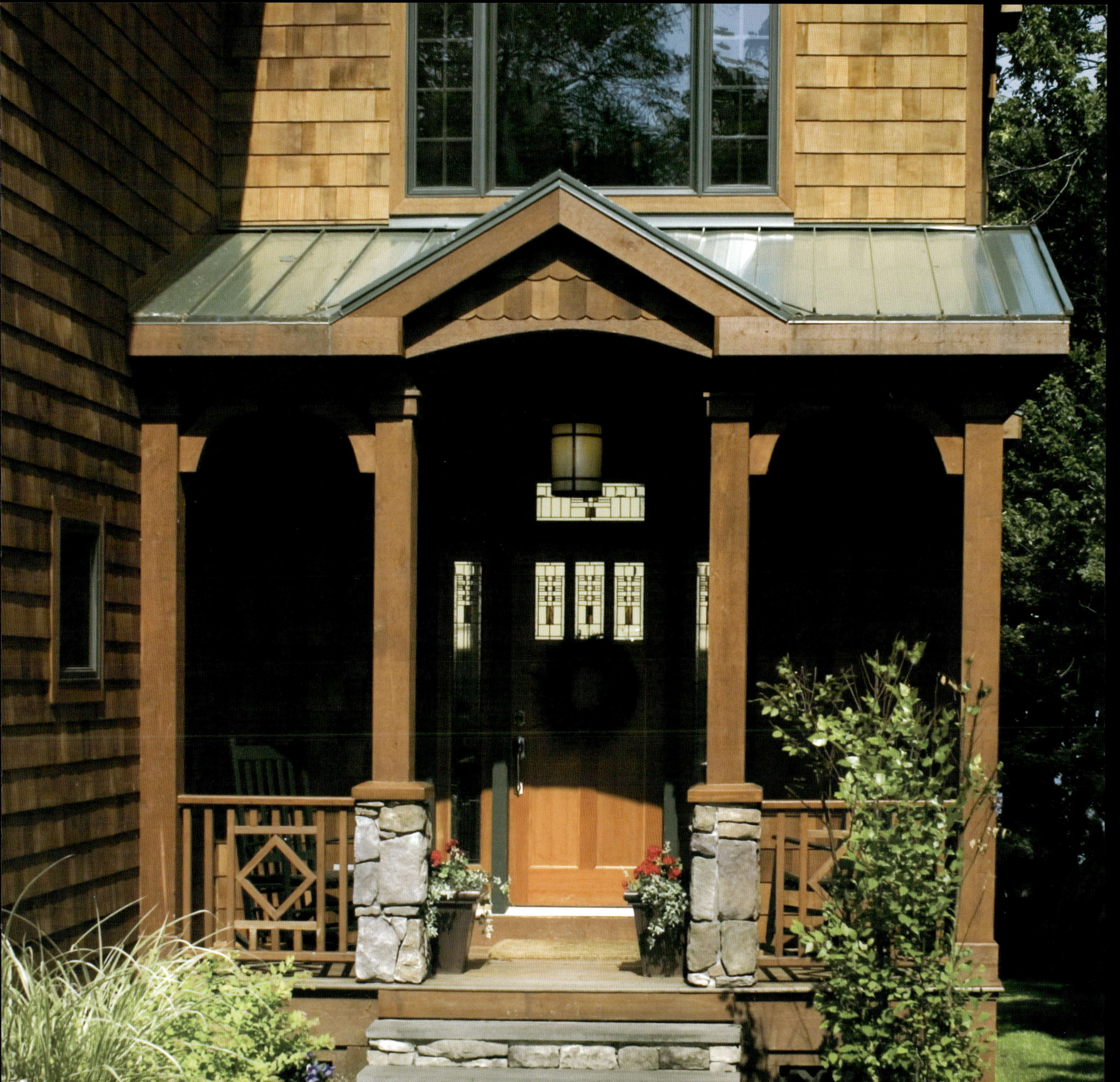

The expansive porch creates an excellent transition between house and lake.

ARCHITECT: *Al Cappelli*
BROKER: *All Homes*
MANUFACTURER: *Haven Homes*
BUILDER: *Bob Fisher*
SQUARE FEET: *5,500*
NUMBER OF BOXES: *14*
LOCATION: *Newburgh, NY*

originally tried to discourage Bob, citing the disparity in the type of house he wanted and the simpler houses that were currently built on the lake. Bob's desire to return to his roots won out and Scott graciously helped him put the project together. Webb says that the modular business has changed so much since he began: "The homes were two boxes; that was it." But Bob had a clear idea of the type of house he wanted to build, and it was much more than the two boxes of yesteryear.

Bob and his wife, Lori, did extensive research into the Arts and Crafts architectural movement, reading every book they were able to find and making several visits to the Adirondack area. The Steeles were taken with the natural materials of copper, stone and wood used on

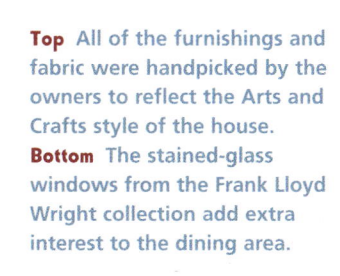

Top All of the furnishings and fabric were handpicked by the owners to reflect the Arts and Crafts style of the house.
Bottom The stained-glass windows from the Frank Lloyd Wright collection add extra interest to the dining area.

"this house was much more than the two boxes of yesteryear"

All of the woodwork and fixtures are consistent with the Arts and Crafts style of the house.

those designs and liked the Stickley furniture that was typical of the period. They read books on Charles and Henry Greene and began to gather items they would like to see in their home. Working with architect Al Cappelli, the Steeles were able to obtain plans for the vision they had

The gas fireplace opens to the bathroom as well as the sitting room. Slate was used as trim to keep all of the materials as natural as possible.

Most of the light fixtures in the house were handmade by artisans.

for their dream house. Typical of the Arts and Crafts period, the house was designed with cedar-shingle siding, overhanging eaves, a stone and tile fireplace, bands of windows, stained-glass windows, beamed ceilings, an open floor plan, natural materials, and a combination of rustic and sophisticated design. The furniture they later selected was simple and functional, with vertical and elongated forms. Typical of this style is the lack of ornamentation and an emphasis on finely crafted furniture.

The Steeles visited the Haven plant, and Bob said he was sold on modular when he saw how they were manufactured. He liked the fact that they seemed very sturdy, were built on jigs, used screws rather than nails and were closed up by foam insulation in every space.

Through Webb, they purchased their home, which was fourteen boxes, and it was set on beautiful lakefront property after an old cottage at the location had been leveled. A new foundation was poured and a rubber-wall weather barrier was added to further protect the basement. After the house was set, builder Bob Fisher came in to do the extensive interior and exterior finish work.

Bob said that initially "the slope of the land was a problem, but we turned it into a positive when we decided to build a deck." The house was also designed with a double-high ceiling in the family room. When the Steeles saw the house set, they decided that the ceiling was too high for the size of the room. They opted to add a floor on the second level and create a bedroom out of the space above the living room. Again, they turned a potential problem into a positive with the additional living space.

Lori and Bob, with their three children, visit the lake house whenever they have the opportunity. Bob says it is a "great place to come to just decompress." They go waterskiing, fishing and boating in the summer and ice boating in the winter. While boating on the lake, they meet friends and invite them back to their house, and before long, the gathering becomes a party.

The cabinet above the fireplace conceals a flat-screen television.

53

The Douglas Cutler House

ARCHITECT DOUGLAS CUTLER became interested in modular houses when he was considering developing some property, and he researched various options. He started out designing stick-built homes but saw possibilities with modular technology and the customization that was available. Along with his own home, he has designed many other modular houses and homes to be renovated using modular construction.

Douglas' wife, Lauren, sees her husband's legacy as designing houses that people really want to live in because "they can have the house of their dreams on a more limited budget." Initially she would have preferred a more traditional house, but after moving into this open, light-filled house, she says she has grown to love it. While the house was being set and completed, Lauren says neighbors constantly watched the process and couldn't believe this house was modular. She says her house is a good example of what can be done with

Opposite page All of the greenery on the lot is a soft contrast to the angular lines and stark white of the house.

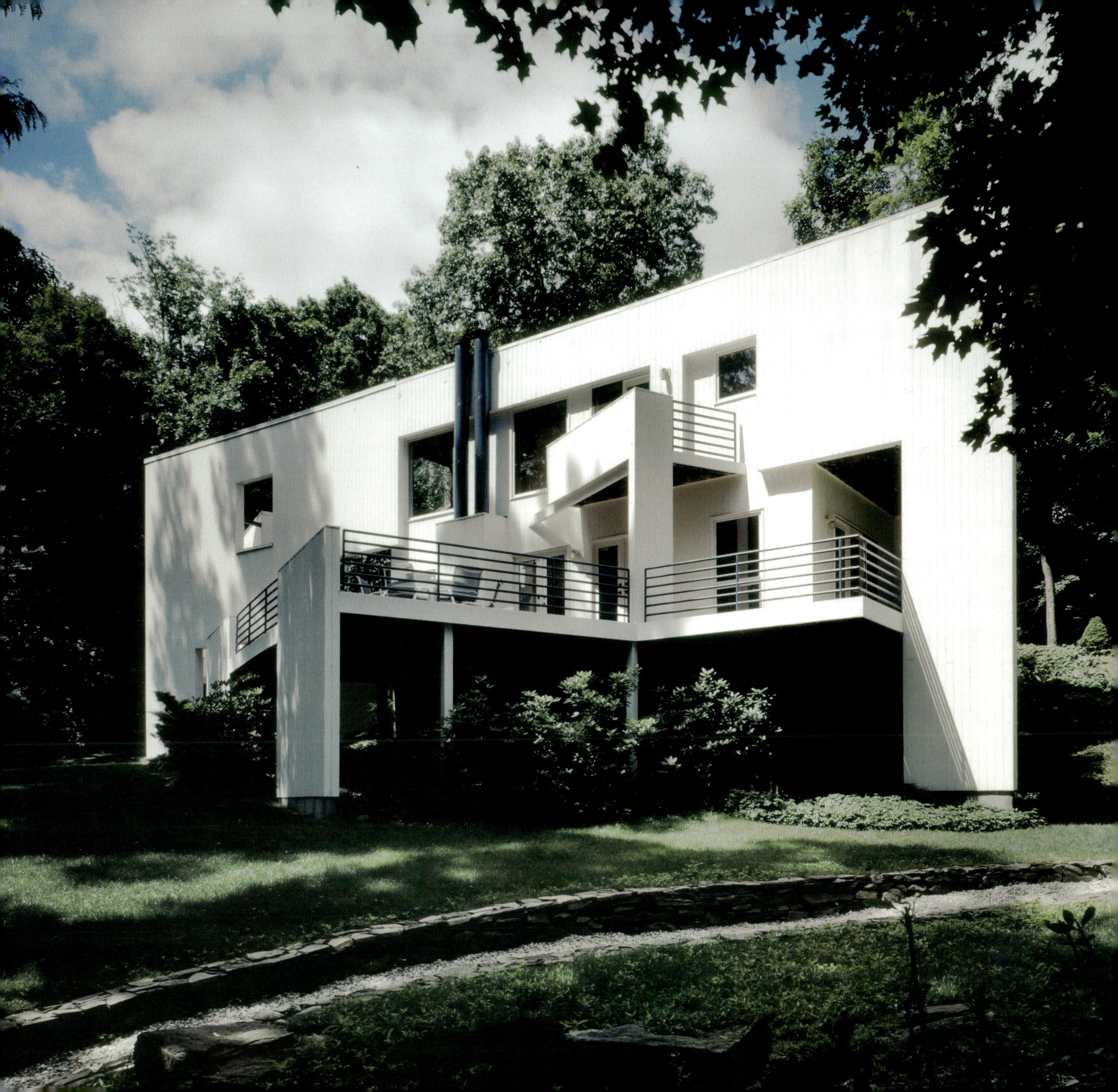

Passersby would be surprised to find out this very modern structure was modular built.

> **Architect:** *Douglas Cutler*
> **Square feet:** *3,200*
> **Number of boxes:** *6*
> **Location:** *Wilton, CT*

modular construction—"[It's] not a style but a type of construction."

Douglas was not deterred from building on this lot in Wilton, even though it had a five-foot-high ledge and wetlands; he opted to use those features as design elements. The rear ledge gives the house privacy, and the deck Douglas put there is a quiet place to sit and read the morning paper. A pond of water is often seen in the wetland area, making the property all the more scenic.

A modified split-level, the house is tied together with custom-made blue metal railings that join the lower living room, kitchen and dining room with the family room on the second floor and the bedrooms on the third level. Douglas stick-built a central area that connects the modules on either side. The blue exterior flue and the blue exterior and interior railings are the only areas of color, and become design features of the house.

The two wings of the house are connected by a glass enclosure made up of twenty-three fixed and stepped panels of glass designed by the owner/architect.

Massive windows unite the greenery with the breakfast area.

Architect Douglas Cutler designed all of the furniture in the house and had it built by a local craftsman.

The owner designed the iron stair railings that connect the three levels of the house.

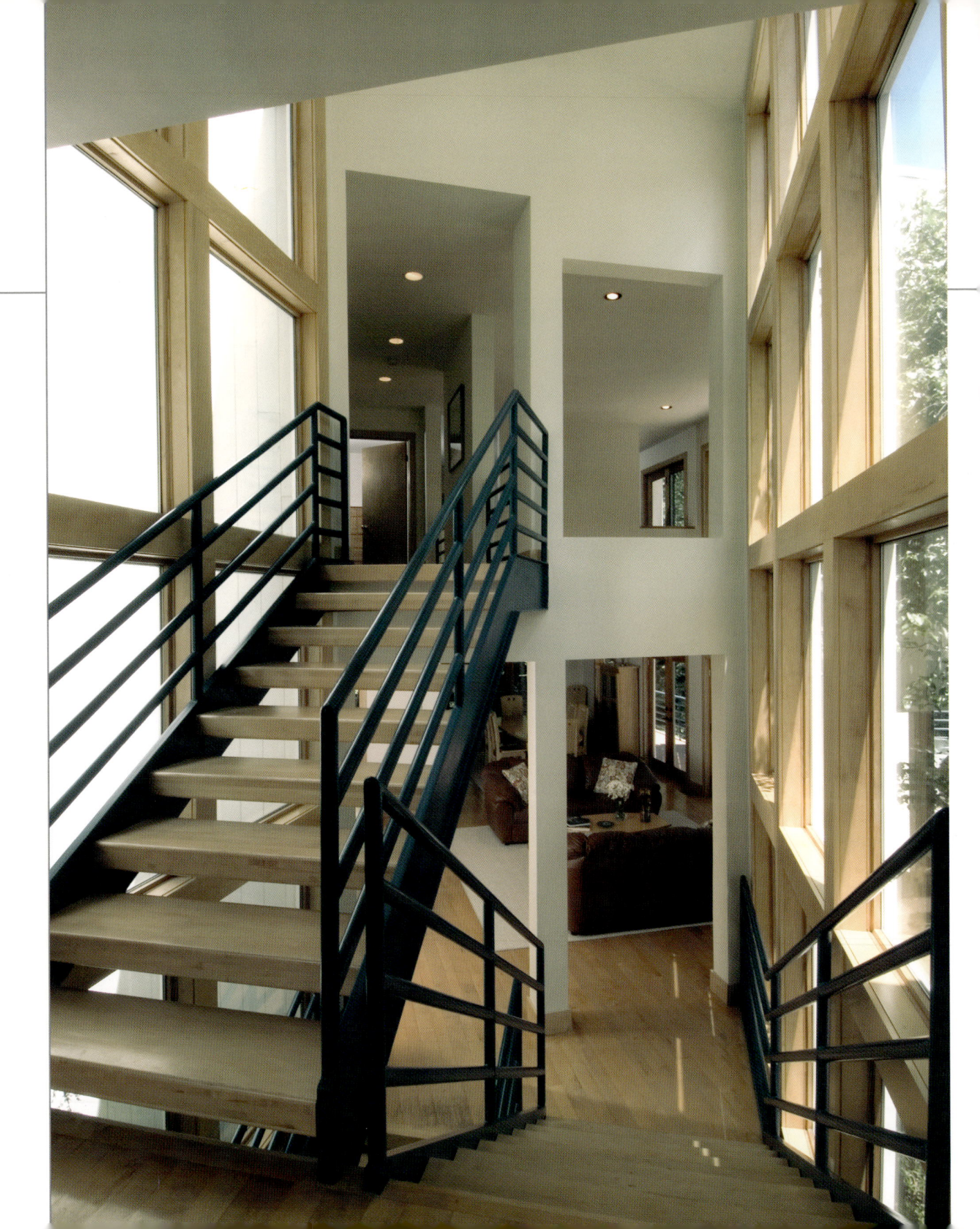

Cutler built the children's tree house himself but says he "wishes it could have been built modular."

Massive windows in the living room and a double-high ceiling create a very light and airy feeling. Double French doors open to a large deck used for extending the living space and entertaining.

There are also several decks on the upper level of the house that extend the living space and open the home to the exterior in many areas. The massive trees and many plants surrounding the house are a colorful contrast to this very angular, white structure.

Douglas designed all of the furnishings in the house, which were built by a local craftsman. The furniture has clean lines and is very monochromatic, creating a perfect synergy with the design of the house. Floors and trim are maple and add a natural color to the stark white of the interior walls. The kitchen cabinets were installed by the modular manufacturer, but the Cutlers opted to have granite countertops installed on-site.

"As an architect, I like to think out of the box and explore all of the new technologies, taking advantage of the labor and time savings offered with modular construction," says Douglas.

The Cape House

ALTHOUGH THIS BEAUTIFUL, recently built house is an imposing twelve thousand square feet, it has the cozy feeling of a cottage that has been there for many years. Built as a family vacation house, it is both functional and beautiful, with gorgeous views of the ocean from many of the rooms.

After living in the original house on the property for three years, the owners decided to level it and put up a more suitable house. They opted to build their new home modular because they felt the workmanship was more controlled and the materials would endure less weathering in the protected factory environment. The owners spent time looking for an architect, modular company and builder. Their observation on the process: "If you are going to build modular, you need to spend a lot of preplanning time to get what you want."

The owners worked with architect Larry Link to create a house that captured the look of a traditional house and incorporated all of their likes and needs.

Opposite page This Georgian revival house has all the elegance of an old-world mansion but with new products to make it practical in its proximity to the ocean. Designed for its island location, all exterior trim is synthetic, including the columns, fascias, frieze and soffits.

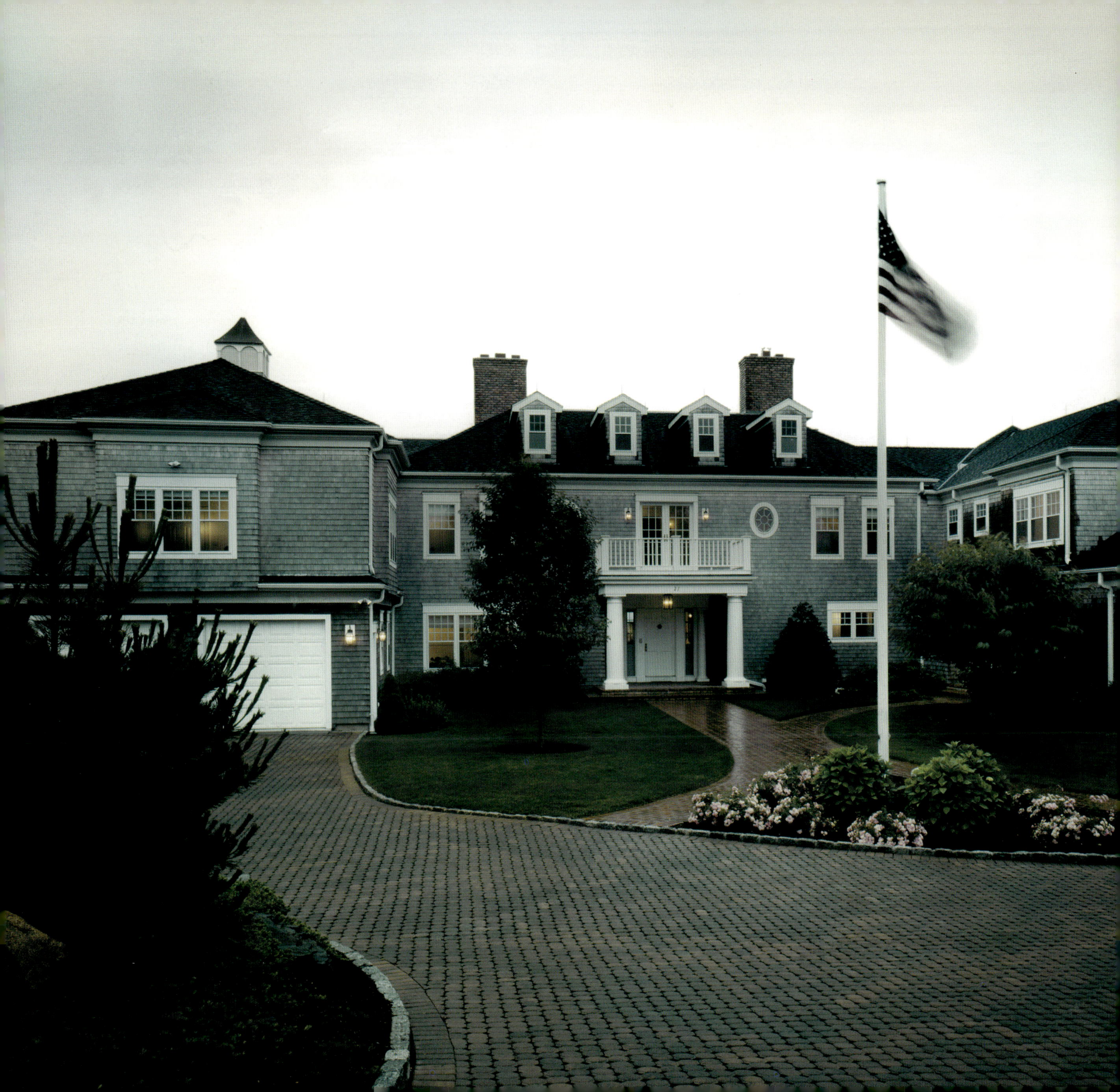

Right Ocean views can be seen while one enjoys a casual meal in the kitchen.
Below Even on a rainy day, this house appears warm and inviting. The multiple columns give the house an elegant look, and the wraparound rear porch not only opens the rooms to the outside, but also adds architectural interest.

ARCHITECT: *Larry Link, L. J. Link Jr., Inc.*
MANUFACTURER: *Epoch Homes*
BUILDER: *Darrell Hoss Builders*
SQUARE FEET: *12,700*
NUMBER OF BOXES: *26*
LOCATION: *Cape Cod, MA*

The owners said they wanted to "create the feeling of a period house." Link, they felt, was very knowledgeable about the different historic periods and would be able to transform their ideas into a "new antique" design that had all the warmth and character they desired. Link wanted every room to be a surprise so that you would have to enter each to see what was there, creating an enticement to move throughout the house.

The owners chose to use as many synthetics as possible in the construction because of the close proximity to the ocean. The windows and doors by Acadia are vinyl to weather the salty ocean air. Many of the abundant moldings are also synthetic for durability. The wood floors, however, came out of an old Pennsylvanian barn from the late 1800s. The wood looked very raw when it arrived, but it was sanded and stained and now gives the house a warm, traditional look.

Top The generous use of molding, the traditional decor and the large fireplace add great warmth to this family gathering place.
Bottom The barbecue required a special venting system to satisfy code requirements. The wok stove is used to accommodate the large lobster pot, which gets much use with this extended family.

Right top The kitchen was designed to accommodate several cooks working together at the same time.
Right bottom Living-room furniture is massive and comfortable for casual family time. The french doors open the living area to beautiful landscaping and ocean views.

Because there are so many diverse eating habits within the extended family, they built an extra-large kitchen so everyone could cook what they like together at the same time. An outdoor grill was adapted for the interior, requiring a restaurant-type fan to meet the local building code. A wok cooker was also installed to be used with a massive pot for cooking lobsters and other shellfish. It is a very workable kitchen with every convenience, including pot fillers, multiple refrigerators and dishwashers, warming drawer, compactors and water purifiers.

In addition to the many windows in the house, decks were built off several rooms to take advantage not only of the view, but the cool, salty breezes. Operable sidelights were used with the front door so a cross flow of air could move through the house, decreasing the need for air-conditioning.

The whole family enjoys vacations there together, reading, fishing, sailing and cooking. The cape cottage is a wonderful place for the generations to come together and enjoy each others' company.

"they wanted to create the feeling of a period house"

Many of the furnishings in the house, such as the large mirror in the living room, were preowned by the family and fit nicely into this oceanside home.

Summerside Cottage

DESIGNED AS A vacation house, Summerside Cottage is close to the Muskoka area of Ontario, north of Toronto. It has large, open interior spaces, vaulted ceilings and a steep roof pitch, an engineering feat in modular construction. This house, like many of the modular homes in Canada, was built by the manufacturer from start to finish, with the exception of the foundation and masonry cladding. In the United States modular houses are predominately sold by builder-dealers, who are the customer's only resource throughout the construction process.

Howard Sher of Quality Engineered Homes says, "We are seeing more one-stories, which I think is a sign of the aging population and also the movement of retirees to rural areas. The trend with retirement homes is smaller houses, bigger upgrades and more luxuries." Although this house is a two-story designed as a vacation or country home, it still has the master bedroom on the main floor, making it appropriate for the owners

Opposite page Built as a vacation cottage, this modular home has an open floor plan and an accessible porch for outdoor entertaining.

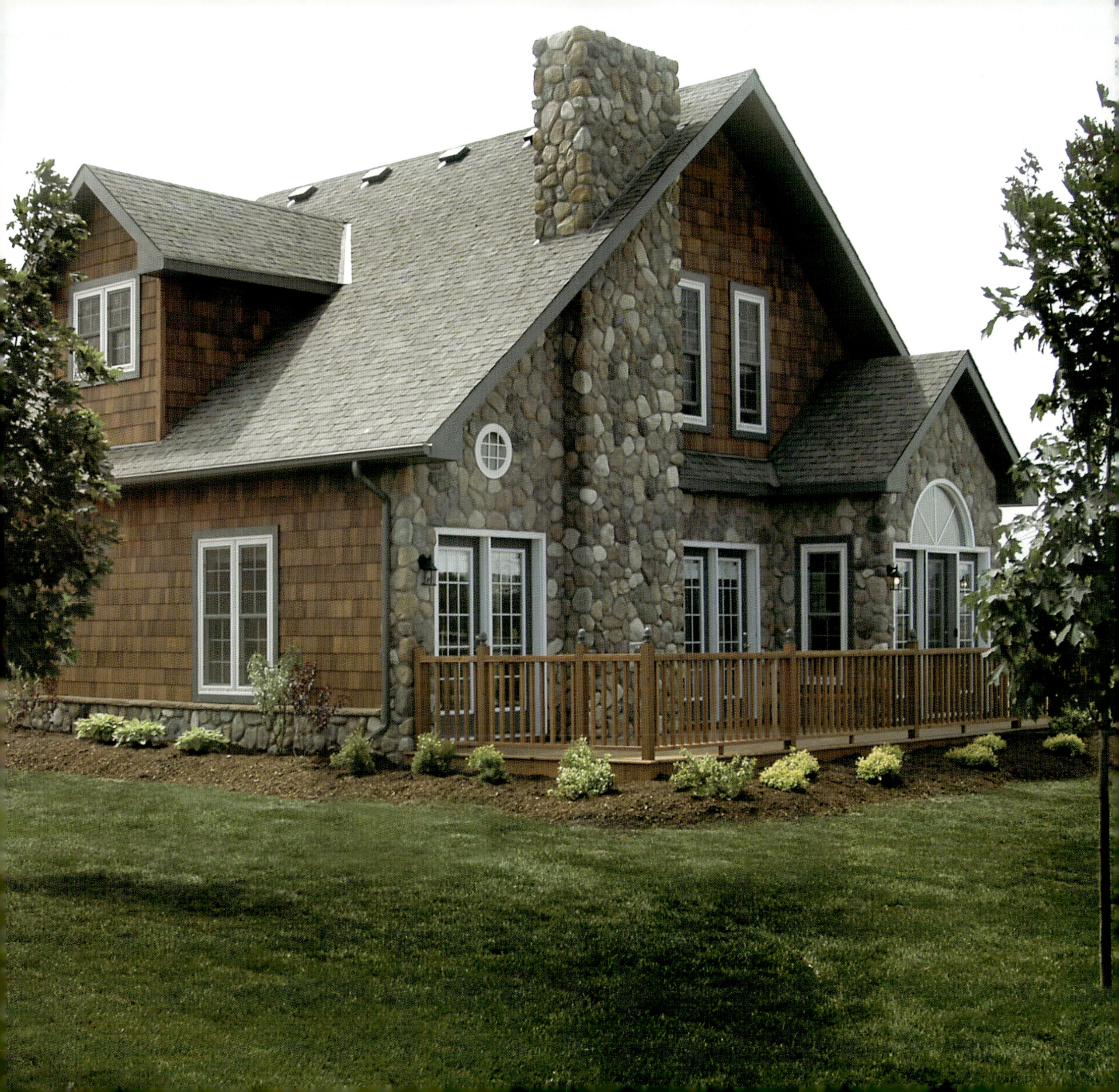

Above The french doors and transom window bring natural light and warmth to the kitchen and breakfast area.
Opposite page A distressed barn beam serves as a mantel on this cultured-stone gas fireplace.

MANUFACTURER: *Quality Engineered Homes, Ltd.*
SQUARE FEET: *1,653*
NUMBER OF BOXES: *7*
LOCATION: *Cookstown, ON, Canada*

to use as a retirement home in the future.

The siding is cedar shakes that have been pre-mounted on plywood sheets (Craftsman panels by Shakertown), which install in about one-third of the time it would take to install individual shakes. The stone veneer is cultured stone, which looks very natural even upon close inspection. The flooring throughout the house is engineered wood (Mirage) with a thick prefinished oak veneer. This flooring has a thick wear layer that can be sanded and refinished, and carries a twenty-five-year warranty. Engineered floors are said to be slightly more resistant to moisture than solid wood, which makes it particularly practical in this lake area.

Canada's Office of Energy Efficiency (OEE) offers the R2000 program to encourage the building of energy-efficient houses that are environmentally friendly. Houses that meet these stringent requirements are

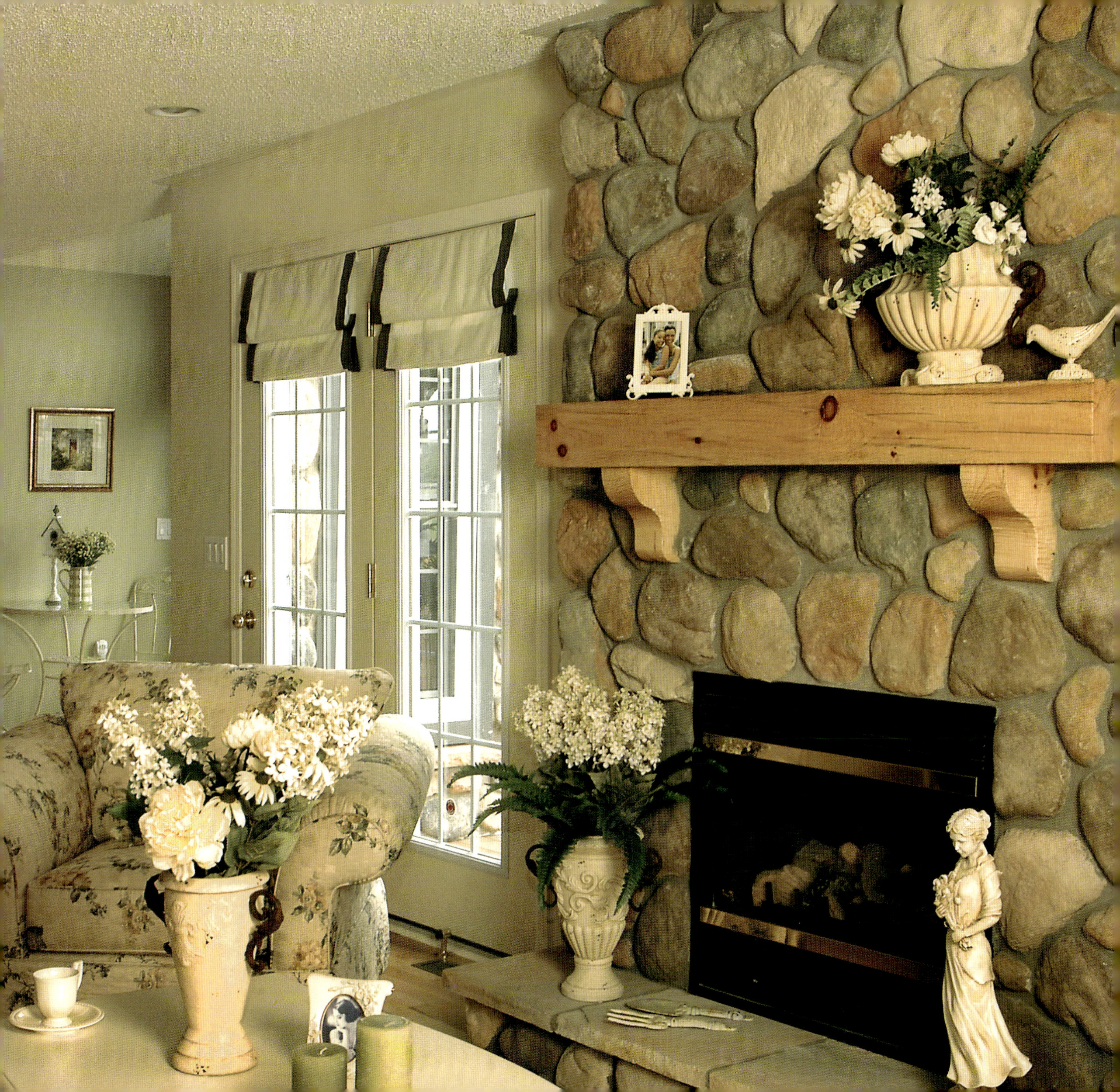

The cabinets in the kitchen are maple veneer, but the doors are solid maple.

about 30 percent more energy efficient than standard homes built to code, and receive a certificate that guarantees these standards have been met. Quality Engineered Homes builds to these standards, so their houses are very well insulated and are less expensive to heat. However, because they are so tight, mechanical ventilation systems are recommended to assure the interior air is circulated and will overcome the buildup of moisture, odors and pollutants in the house. A heat-recovery ventilator (or air-to-air heat exchanger) was added to the mechanical system of Summerside Cottage to ensure continuous, or timed, ventilation throughout. Interior air is exchanged with fresh exterior air while the heated or air-conditioned air is exchanged at the same time. This results in lower utility costs and a more comfortable house.

This entire house was built to meet stringent energy-efficient standards with the use of materials that ensured the house would hold up over the years and be completed on time and within budget.

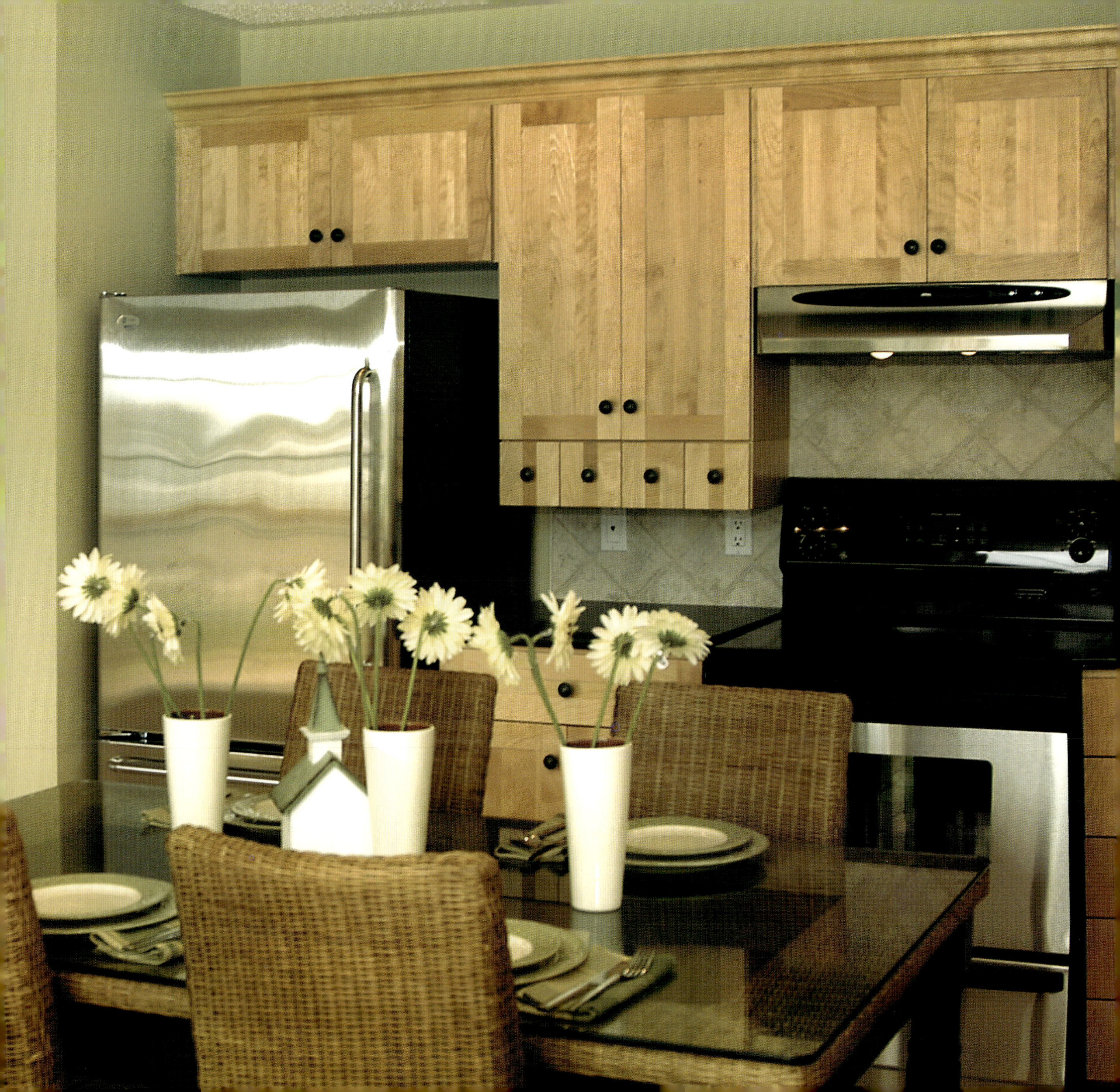

Lake Winnipesaukee

LAKE WINNIPESAUKEE in central New Hampshire has a long Indian history; it is said the area was inhabited by Native Americans for ten thousand years. Legend has it that the children of two hostile tribes on the northern and southern shores of the lake fell in love and wanted to marry. Mineola, the Indian maiden, convinced her father to reconcile with the other tribe and allow her to marry Adiwando. Both tribes were in canoes, seeing the young couple depart after the wedding ceremony, and the sky was very overcast and the water black. Just as Mineola and the young chief Adiwando turned to leave the others, the sun came out and the water sparkled. The girl's father said this was a good omen and named the waters Winnipesaukee, or Smile of the Great Spirit.

Over the years, many Indian artifacts have been recovered in the area, including spearheads, pottery and stone tools.

Opposite page The multitude of windows in Cottage on the Lake were created to take advantage of the beautiful lake views.

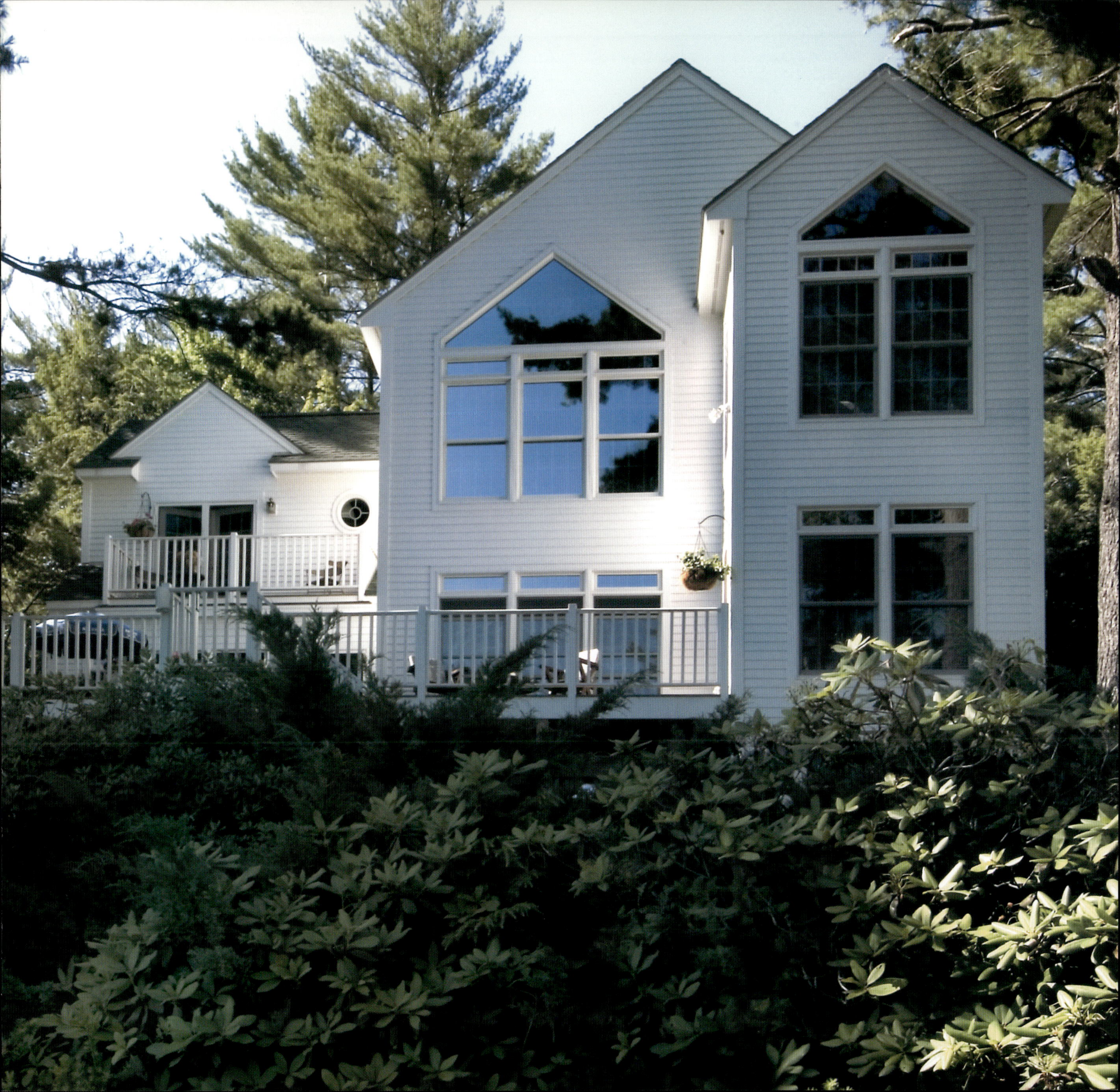

Lake Winnipesaukee
"Smile of the Great Spirit"

The front of the house appears small and belies the home's spacious interior.

The lake is seventy-two square miles and has approximately 274 islands. It is the largest lake in New Hampshire and the sixth largest natural lake completely within the borders of the United States. The third largest island on the lake is Cow Island, which is 522 acres and was once used to quarantine cows brought to this country from Europe. The fourth largest of the islands is Governor's Island, which is 504 acres and is an exclusive community with beautiful homes used both year-round and for vacation.

With the White Mountains as a backdrop, Lake Winnipesaukee is a popular vacation spot for fishing and boating, and offers many family activities. Ice fishing, skating and snowmobiling are popular in the winter, and boating and swimming are popular in the spring and summer. *The Mount Washington* is a 230-foot cruise boat that tours the lake six months of the year, offering some beautiful views of the lake shores.

COTTAGE ON THE LAKE

MANUFACTURER: *Epoch Modular Homes*
BUILDER: *The Fournier Group*
SQUARE FEET: 2,470
NUMBER OF BOXES: 8
LOCATION: *Meredith, NH*

Prior to the construction of this beautiful home, there was a small, one-story house on piers on the property, which the current owners had to demolish to set their new

The terrace off the master bedroom is a wonderful place to read and enjoy the tranquil peacefulness of the lake. The vaulted ceiling and interesting window configuration add to the charm of the room.

The configuration of windows takes advantage of the lake view and drenches the room with sun.

The high ceilings, abundant windows and open floor plan give this house an airy feel.

home. Because of local ordinances, the new cottage had to fit within the footprint of the old house. In order to achieve more space, the owners opted to build the house with three floors, expanding the square footage without expanding the size of the footprint. They designed the lowest level as a play area for the family and installed a bathroom that could be used for showering after a swim in the lake. The middle level was designed for the common areas and the upper level is where the bedrooms are located. A terrace off the master bedroom was designed as a place to sit and have tea in the morning or a glass of champagne in the evening.

Small boxes had to be used for this site because of the narrow dirt roads leading to the house. It would have been impossible to bring in the larger boxes used for more approachable locations.

The owners of this cozy home requested numerous and varied windows be used on the lake side of the

Large double-hung windows with transoms in the living room shower this charming cottage with light and beautiful views of the lake. The windows are Low-E glass with argon-gas inserts, which help insulate the house and prevent discoloration of fabric.

The style of this boathouse on the dock is consistent with the house design. Since boating is a central activity for the owners, the boathouse serves as a convenient place to store equipment.

house to take advantage of the beautiful views.

They were also concerned that their new home would be out of proportion to the other houses in the nearby area. The front of the house was therefore designed to look like a small cottage, while the rear of the house was given a more sprawling appearance.

Construction on another house, a stick-built one, in the cove was beginning at the same time as the cottage. Since that winter was very cold and snowy, it took a year for the builders to complete the house across the way. In contrast, the modular house was constructed in a controlled environment, so there was no delay in completing the house, and it was set in the early spring.

LAKE WINNIPESAUKEE

79

Although this house has a simple front facade, other than the multiple dormers, it hides an exciting and expansive house behind it.

Governor's Island Paradise

Manufacturer: *Epoch Modular Homes*
Builder: *The Fournier Group*
Square feet: *4,500*
Number of boxes: *6*
Location: *Gilford, NH*

Though it was originally built as a vacation house, owners Ann and Kevin Attar say they like their Governor's Island house so much that they hope to make it their permanent residence in the near future. Kevin initially opted for a modular house because it was planned to be a vacation home, which he thought would get limited use. He felt building it modular would also enable him to use it more quickly. If it was site-

The backyard area is beautifully landscaped and includes a three-tiered waterfall and a sitting area for enjoying the view of the lake.

built, Kevin was concerned about the elements having a negative effect on the construction materials since the house was so close to the lake.

While the Attar house was under construction, Kevin and his father visited the modular factory to see the progress on his home. "We walked into the kitchen, which was the smallest box, and there were nine people working; then we walked into the master bedroom and there were eleven people working. The amount of manpower concentrated in each of the units at any given time was very impressive."

The Attars planned to demolish the old house that was on the property

Opposite page The double-high ceilings and open floor plan give the common areas a very airy feeling.
Below The master suite opens to the gracious deck and beautiful views of the lake.

and designed the new house to fit on the old foundation. But, just before the boxes were ready to be set, the foundation collapsed. They were forced to dig out the old foundation and construct a new one. They were relieved, however, that the foundation collapsed before the house was set rather than after, when it could have been disastrous. Although it was an unfortunate occurrence, it gave the Attars the opportunity to set the house back twenty feet and put a larger garage in the front of the house. Their only regret was the two-month delay caused by the construction of the new foundation.

The floor plan was designed to be very open and spacious with a double-high ceiling in the living room, giving the house a particularly airy feel.

The Attars added many custom items to their home, including a large deck made of composite decking, which requires minimal maintenance, won't splinter and is softer to walk on than wood. They built a second garage for their lake-area "toys"—the snowmobiles, jet skis and so on. They also added a steam shower, Jacuzzi, hardwood floors, columns and upgraded counters.

The property, which rolls down to the lake, is landscaped with lush gardens and three-tiered waterfalls, which add to the magnificent view from the rear of the house.

With their beautiful, spacious home surrounded by the tranquil waters of Lake Winnipesaukee, it is no wonder that the Attars decided to make this their permanent residence.

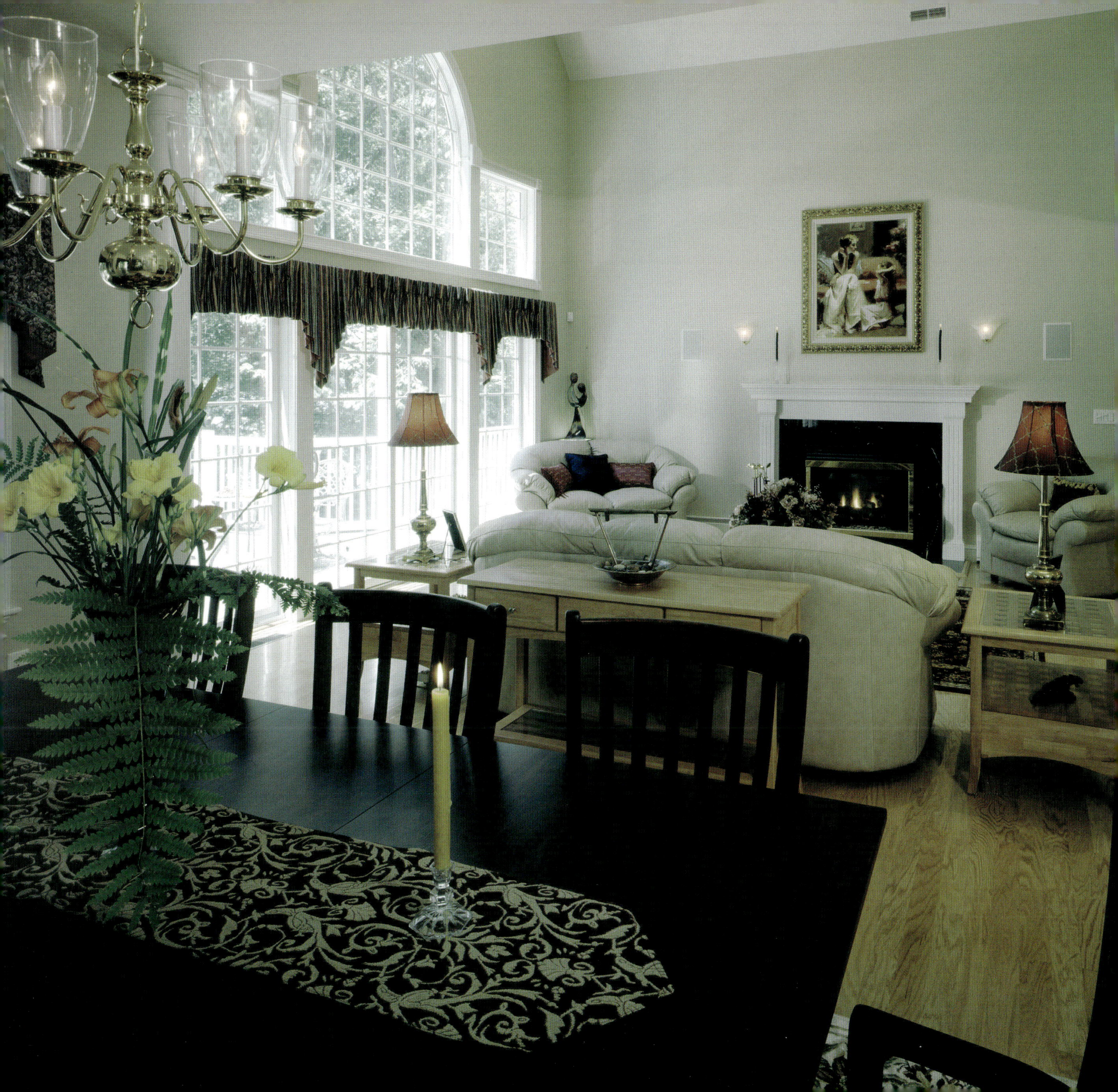

Batkins' Arts & Crafts

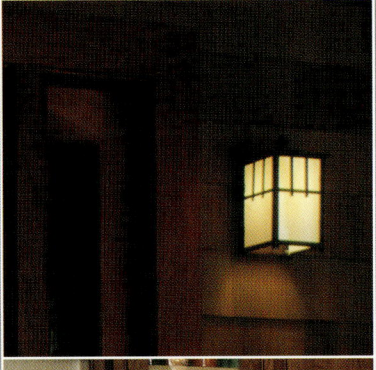

ANNETTE AND STEPHEN BATKIN owned a beautiful Victorian house several miles out of the center of Greenwich, Connecticut, which they had lived in for fourteen years. They decided that since their children were grown up, they wanted the convenience of being able to walk into town. They felt modular construction would be best suited for the small size of their lot and the type of house they envisioned.

Annette says, "Steve and I are very practical people and saw modular as a practical approach to solving our housing needs." They chose their modular company by speaking with friends who had successfully built one of their houses in town, and from positive feedback they received in an online chat room about modular houses. Annette says that they learned a great deal about construction, negotiating with contractors, finding out about available materials and building a better-quality house. "We did our homework

Opposite page The cedar siding was installed on-site, as it is on many larger modular houses, because of the difficulty of lining up the siding on multiple intersecting units.

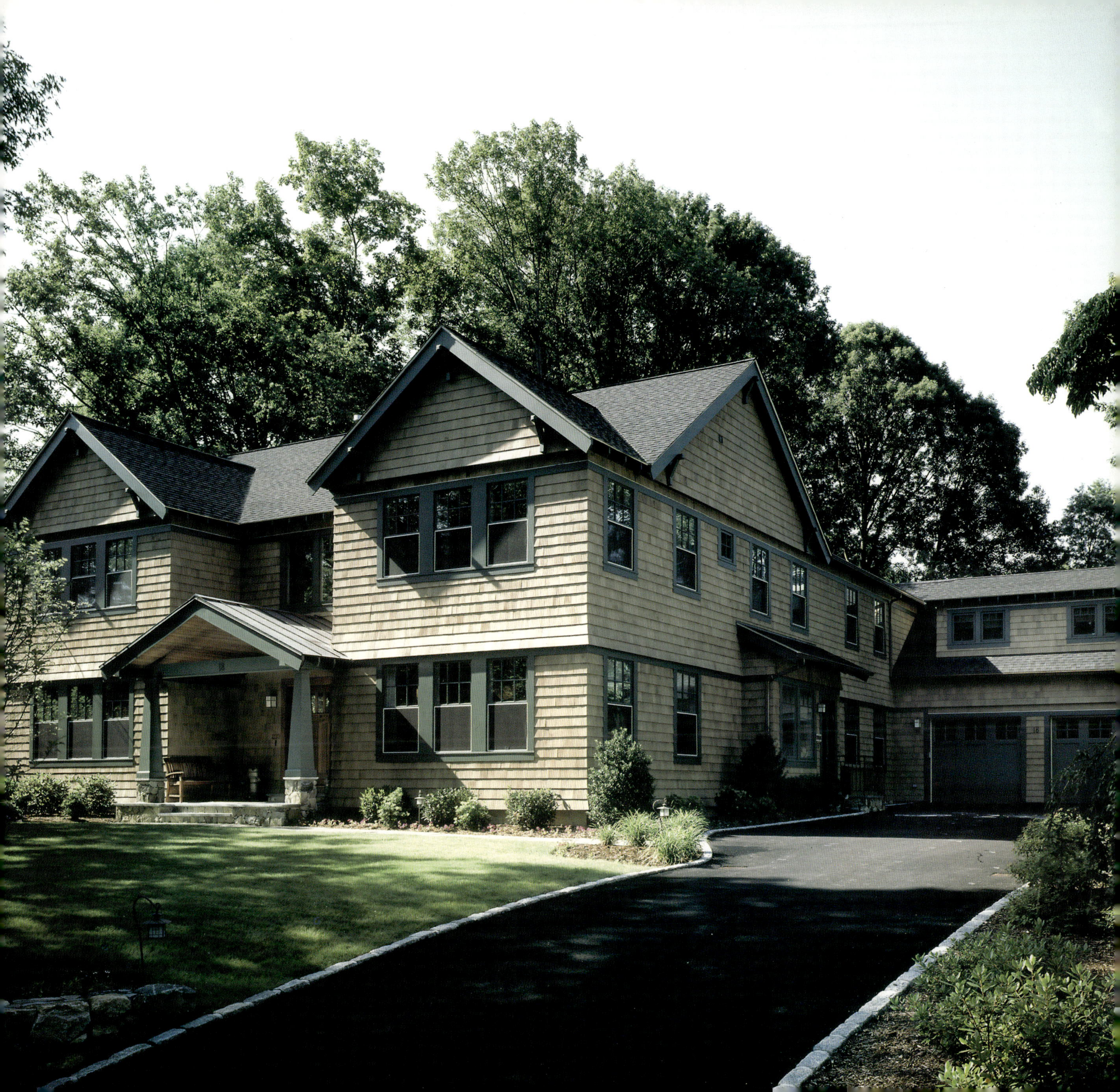

The Batkins took advantage of the high rear ledge by adding a fishpond and waterfall. The landscape adds privacy and a sense of seclusion on an otherwise small lot. The outdoor kitchen has a cooktop, refrigerator and barbecue, making entertaining as easy outdoors as it is inside the house.

Manufacturer: *Epoch Modular Homes*
Builder: *Ed Sandor*
Square feet: *6,500*
Number of boxes: *14*
Location: *Greenwich, CT*

first, so we were ready to go when we closed on the property," Annette said.

With the large size of the house and the small lot size, the Batkins tried to take advantage of the surrounding property. They chose to turn a steep rear ledge into an asset by creating an architectural detail with a waterfall, koi pond, plants and a rock garden. While they were excavating for the new house, they had the builders create a steep slope in the back of the house that would be perfect for a waterfall. Annette had always wanted a pond, so they worked with the landscapers to create a system with the water flowing into the pond and recirculating up the hill. They saved the rocks and boulders that were dug up, and used them to create an environment in the back that is both visually attractive and low maintenance. Typical of

A slate fireplace is the focus of this comfortable sitting room.

Arts and Crafts houses, there is a synergy here between the architecture and nature. The Batkins decided to build a more casual Arts and Crafts/Adirondack-style house because they felt it would fit comfortably into the surrounding neighborhood. Like many of the houses of that period, its siding is cedar shingles, it has bracketed overhanging eaves and there are bands of windows connecting the house with the exterior. The stained-glass window and the natural slate fireplace in the family room are also typical of the era. The furnishings are boxy and simple, with many wood slats in rich wood tones. With an abundant display of handcrafted

Family crafts are displayed at the entrance to the game room.

objects, the Arts and Crafts style is well represented in the Batkin house, with many of the art objects created by Annette and her son Aaron.

While living in the Victorian house, Annette had collected family heirlooms consistent with that period, including cut-glass pitchers, china and so on, which she now displays in a case in her dining room. But while the family lived in a rental house, waiting for the completion of their new house, Annette had plenty of time to start a whole new type of collection, reflecting

The Batkins were able to find a stock staircase that worked well with the Arts and Crafts style of the house.

the shift to an Arts and Crafts style. Using eBay as a resource, she started collecting teapots from the 1920s to the 1950s. She became familiar with the manufacturers of that time and selected pieces from those that would look good displayed together. She later started looking for canisters from the same period that also could be displayed in her new kitchen. The Batkins designed their kitchen both to display Annette's new collections and to be as functional as possible.

Right Many family heirlooms are displayed in this subdued family dining room.
Opposite page The Batkins found an Arts and Crafts door design they liked, and then found a craftsman on the Internet to make it for them.

Not only did the Batkins design their house, but they decorated it as well. "When it comes to building your own home, it is a lot of fun creating it, but the difficult part is having to do everything at once," Annette says. Choosing doors, hardware, flooring and so on, all at the same time, was difficult. She says, however, that working with her husband, Stephen, made for a great blend of talents; Annette's background is in design, so she chose many of the items for their aesthetic appeal, and Stephen, who has an engineering background, got the final say from a mechanical standpoint.

Annette found vintage-type fabric on the Internet and sewed her own

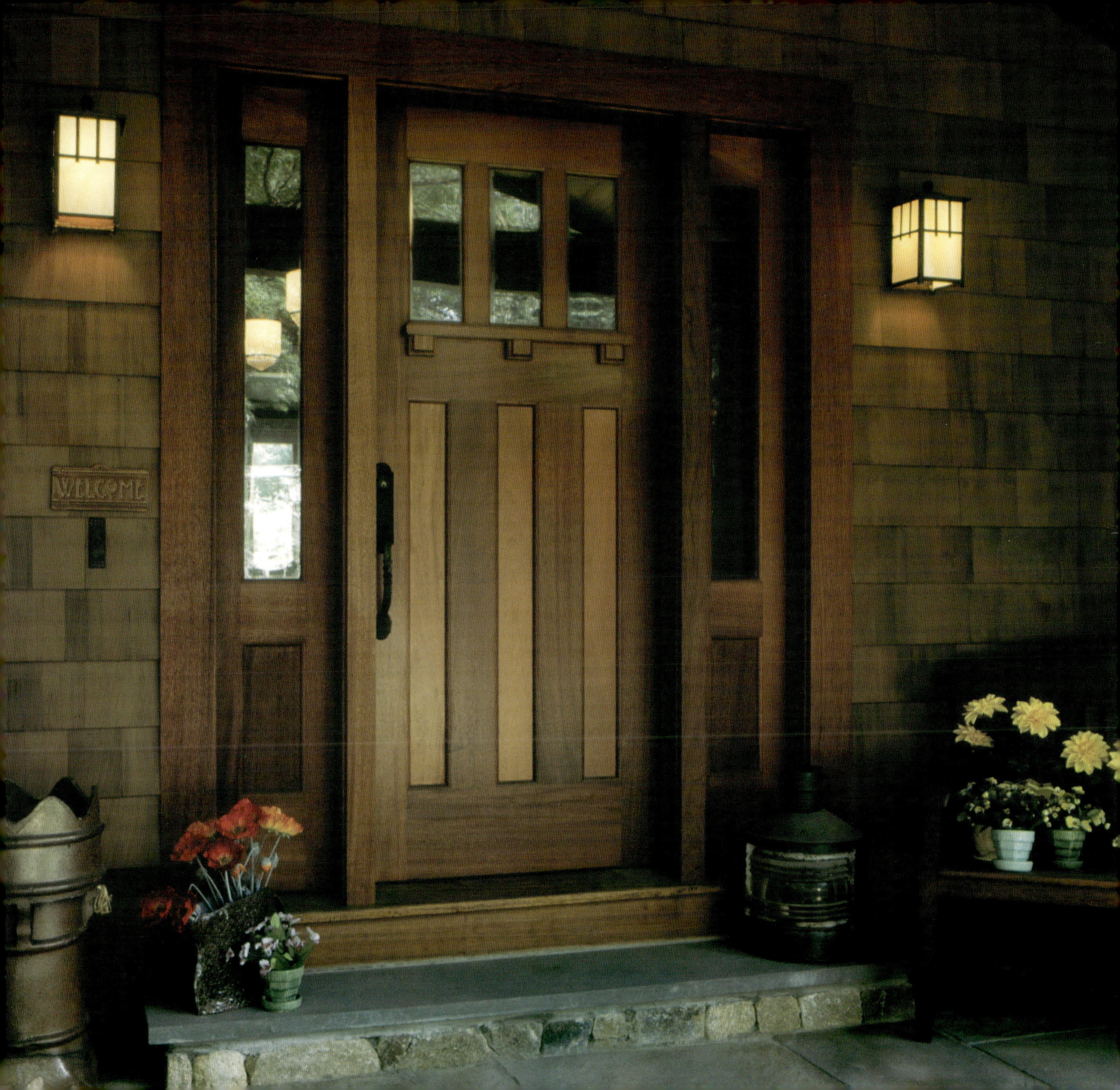
WELCOME

Right The french doors in the living room open to the outside gardens and sitting area.
Opposite page Collected antique canisters and teapots decorate this country kitchen.

window treatments and pillows, which are in abundance around the house. The Batkins' son Aaron, who has a terrific talent for pottery, has added to the beautiful collections around the house with his raku pitchers and other handmade pieces displayed in the living room. Their other son, Adam, helped to create a computer network that would work well for the family.

The Batkins do a great deal of entertaining, and they have created a home that has great warmth and is reflective of the family's collective talents. Annette commented that she loves her house and that one of the great features is "being able to decide to go to the movies at the last minute, walk to the theater and be there just on time."

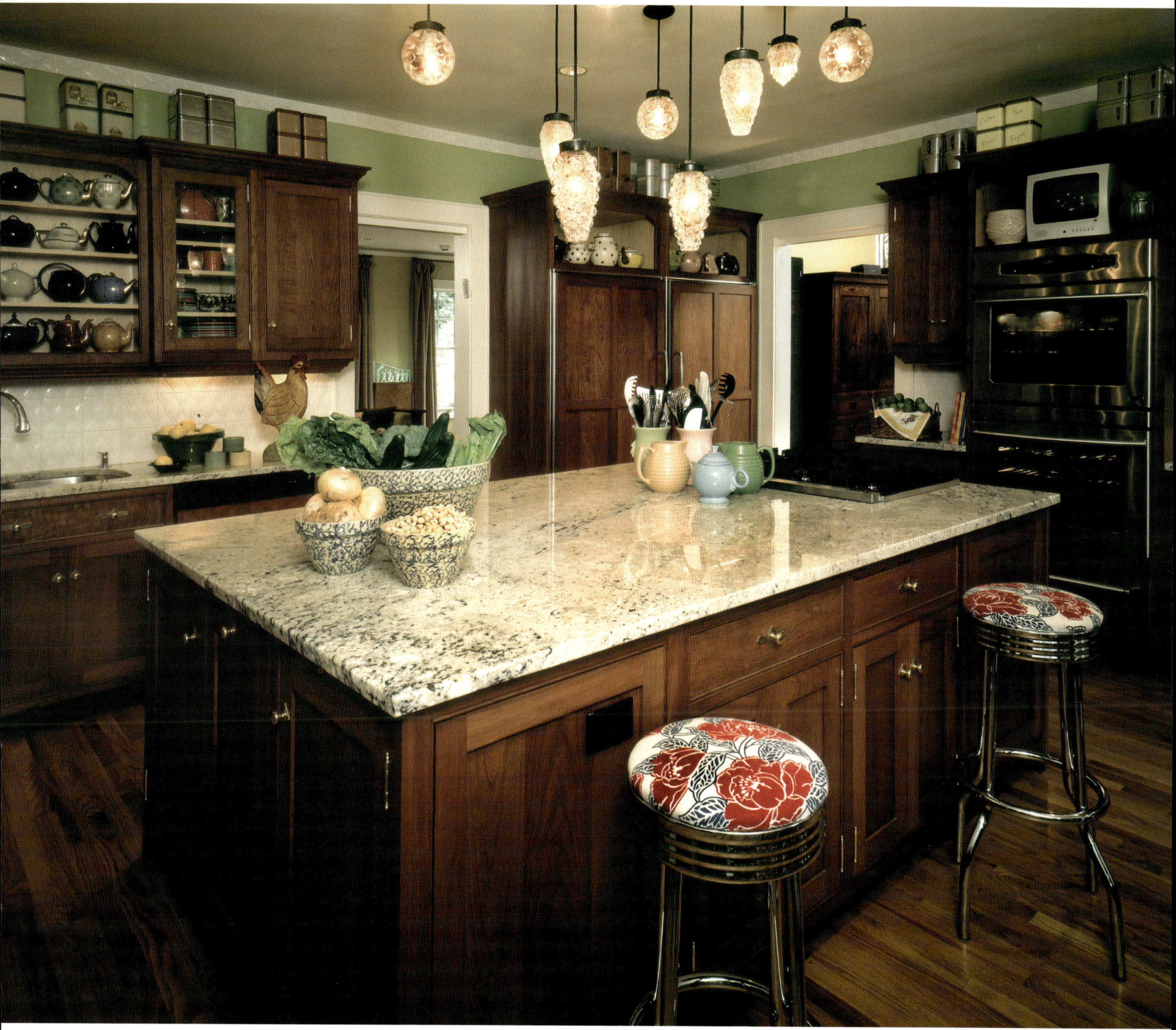

Fanelli Homestead

THE FANELLI HOME can most easily be described as generous in scale and design. Gail and Bob Fanelli built their house to accommodate themselves, four grown children, children's spouses, grandchildren, and other relatives and friends. Avid entertainers, the Fanellis had both the interior and the beautiful, immense gazebo outside designed for large groups to gather. At 6 feet 5 inches, Bob is a big guy and he wanted hallways and doors to be generous. Ceilings are 10 feet high on the main floor and doors are 7 feet high (larger than the traditional 6-foot, 8-inch size). Vanities in the master bathroom were built at different heights to fit Bob's tall frame as well as Gail's more petite size.

Upon entering the Fanelli home, a feeling of tranquility comes from hearing what one would believe are the sounds of a very accomplished pianist playing, but after taking a quick look, one realizes it is a player piano, run by a computer built into the baby grand. A variety of music, from classical to popular tunes, is

Opposite page The Fanellis wanted their home to have a Colonial look with a lot of symmetry and all spaces on a generous scale.

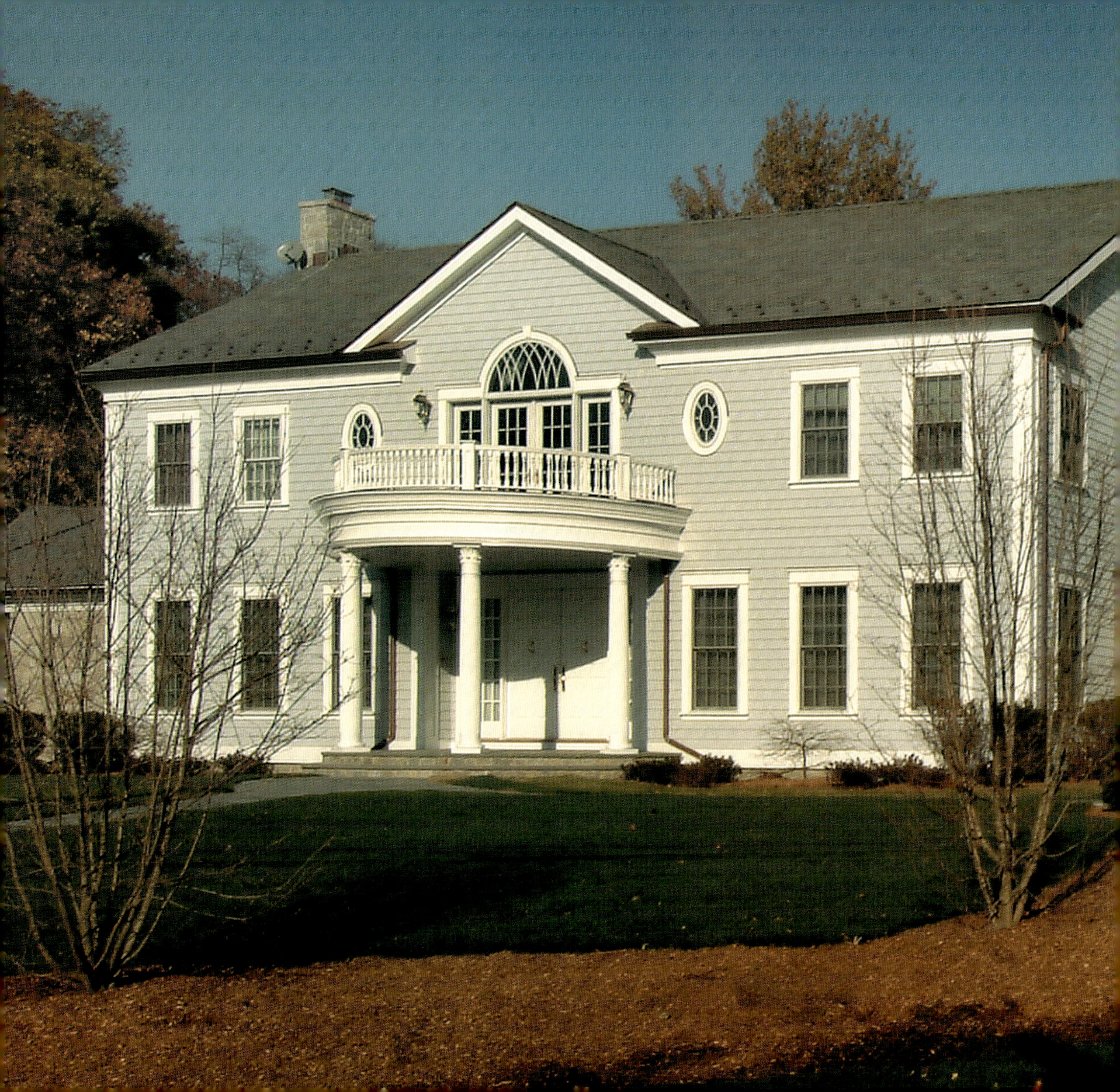

Below left The massive chandelier, made of Swarovski crystal, can be mechanically lowered for cleaning.
Below right The trusses in the kitchen add warmth to this large, open space.

ARCHITECT: *Porter Clapp Architects*
MANUFACTURER: *Customized Structures, Inc.*
BUILDER: *Darrell Hoss Builders*
INTERIOR DESIGNER: *Joe Fanelli Unlimited*
SQUARE FEET: *8,000*
NUMBER OF BOXES: *11, plus 3 panelized sections*
LOCATION: *New Rochelle, NY*

programmed into it, and one cannot feel anything but peaceful in this music-filled house. Bob Fanelli used to sing with a barbershop quartet but

Below left The whole house has an open and airy look.
Below right It always sounds like a piano concert in the Fanelli house, with the player piano going much of the day. The still life beside the piano was painted by Joseph Fanelli.

now joins his family to sing around this beautiful piano at gatherings.

The family was living in a nearby two-story Colonial, but it was becoming more and more difficult for Gail to walk the steps with her advanced arthritis. However, when Gail and her husband found out what it would cost to remodel their new home, they decided to rip it down and start from scratch so they could build the house they truly wanted. With limited lot space, they opted to build the master bedroom on the main floor and to add another floor for guest bedrooms, which could be accessed with an elevator or stairs. A bathtub was sunk into the floor of the master bath; jets, lights and steps were added so Gail could easily get in and out of it. In the

Left The spa tub was sunk into the floor to make access easier. Grab bars were installed in several areas for safety.
Right Doors were installed along the length of the rear of the house to open it up to the immense gazebo outside.

shower area, showerheads were put at various levels to accommodate Bob's height and also the small grandchildren when they stay over.

Very traditional in design, there is a feeling of warmth derived from the beautiful, rich Brazilian Cumaru wood floor and luxurious blue rugs throughout the house. The trusses in the kitchen add a homey rustic touch to a very large, open room. The house's interior was designed by Bob's brother Joe, who is also an artist; his still-life paintings adorn the walls in several rooms of the house.

The gazebo is a wonderful place to relax or to entertain in the warmer months.

The builder, Darrell Hoss, said this house was extremely labor intensive to build because of the intricate moldings, built-ins and details included in the design. Although he is used to building large homes, Darrell said this was the most challenging of the numerous modular houses he has built thus far. The end result, however, is a gorgeous home created by a very cohesive, professional team of architect, builder and interior designer, working together with the homeowners in the friendliest possible manner.

Canyon View Log Home

PEOPLE DREAM OF having a beautiful log home by the lake, with a roaring fire and woodsy appearance. Now that can be achieved faster and less expensively with modular construction without sacrificing the look and comfort of a log cabin.

Canyon View was built with two-by-six lumber, as traditional site-built homes would be, only it was built in a modular factory. The wiring and plumbing are hidden between the walls, unlike conventional log homes in which these items must be routed in keyways between the logs. The half-log siding was applied on-site after the house was set. It is difficult to differentiate the siding on this house from the more traditional full-log system. Half logs are trunks that are split in half and attached flat to the sides of the house. In traditional log construction, the whole log is used and stacked to form the sides of the house.

Opposite page Cultured stone was used on the chimney to give the look of stone while keeping the cost of the house down.

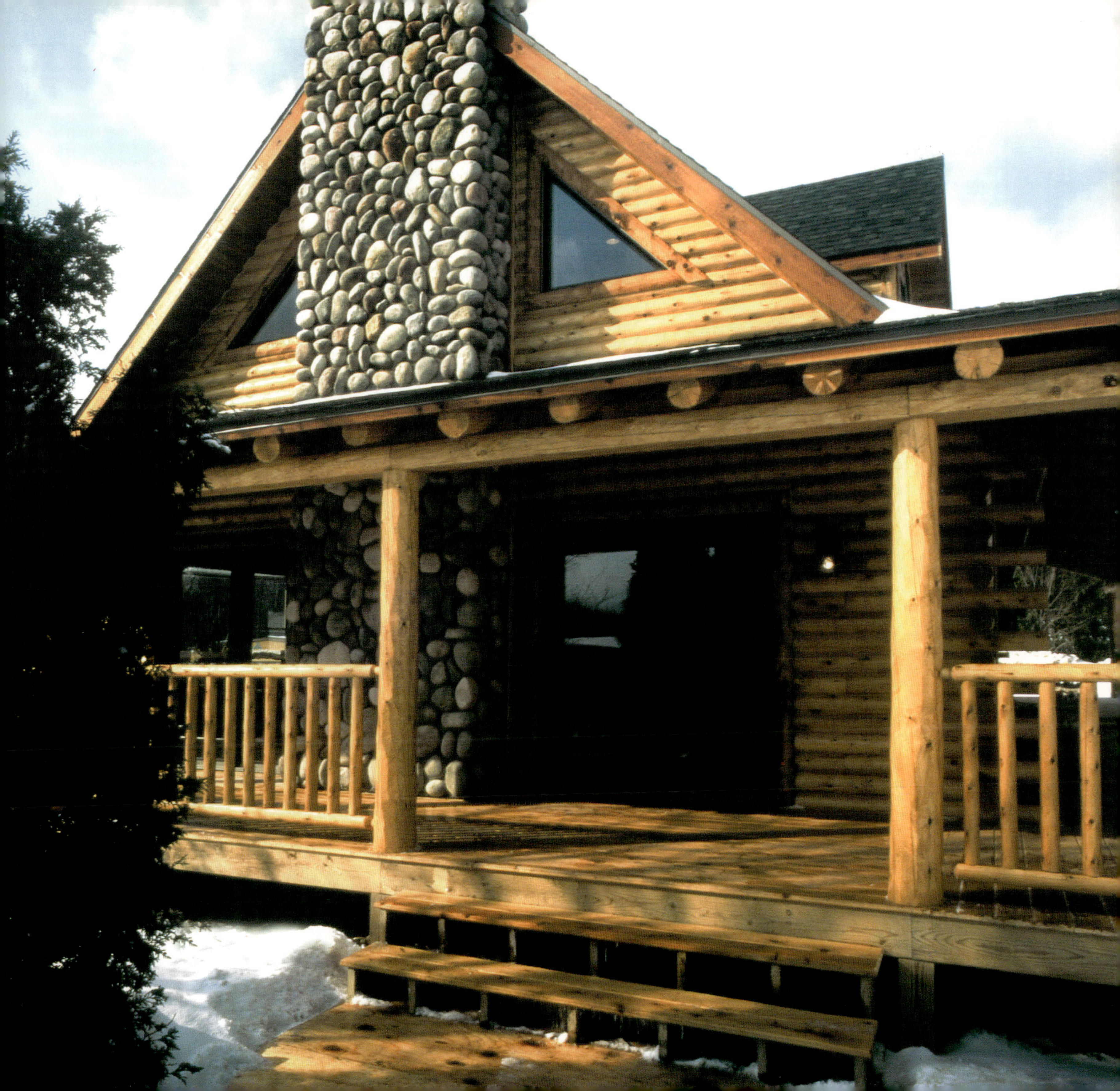

Below The porch is a great place for sitting outside and relaxing. The overhang also protects the house, diverting snow and rain away.
Right The cedar-lined double-high ceiling in the great room provides warmth without the massive trusses traditionally used in log homes, which are very costly to assemble in the factory and reassemble on-site.

MANUFACTURER: *Ameri-Log Homes*
BUILDER: *Town & Country Cedar Homes*
INTERIOR DESIGNER: *Sue Bartlett, Bartlett's Home Interiors*
SQUARE FEET: *2,000*
NUMBER OF BOXES: *2*
LOCATION: *Petoskey, MI*

The cedar trees for this home were farmed in upper Michigan and Canada by the manufacturer. The logs were then milled and kiln-dried in a state-of-the-art mill facility. Because the wood was kiln-dried,

The rustic decor is consistent with the log construction.

there is less chance of it twisting and checking as would be expected with green wood, which gradually shrinks over time as the water is released. The builder chose cedar because of the wood's natural resistance to insects and its tendency to resist water absorption, which can lead to mildew and decay. Only the heartwood part of cedar is used; the outer sapwood rings are removed so as to obtain maximum durability.

Above left Cedar is carried through the house to the bathroom, where it is used to enclose the bathtub.
Above right Hickory wood is used for the kitchen cabinets because it is very hard, and it complements the look of the surrounding cedar.

Like traditional log homes, this house has a butt-and-pass corner design where the logs meet. This is one of the simplest and most commonly used designs in log construction. The logs alternate from layer to layer, so every other log on one wall is a butt and the logs in between are passes.

The interior of the house, including the cathedral ceiling, has tongue-and-groove, end-matched cedar paneling. The staircase is constructed of half logs and the railings are full logs.

One of the most inviting features of this house is the wraparound porch with log rafters. This part of the house was built on-site rather than in the factory. The large overhang provides protection from snow and rain as well as a pleasant place to sit and spend time with family and friends.

The great room has a beautiful cultured-stone fireplace with a cedar mantel, perfect for warming the hands and the spirit.

As is appropriate with this type of house, the decor is rustic and cabin-like in design.

Though still a symbol of the frontier, the new log home can incorporate modern technology without sacrificing the warmth and coziness of log homes of yesteryear.

The newel posts and banisters are made of full logs, while the staircase itself is constructed with half logs.

The Georgian

MARILYN BROWN KNEW about modular houses from her short stint as a real estate broker. While living in a fixer-upper she had remodeled, Marilyn decided her next house would be a modular because, she says, modulars are "better built, better value and take less time." When Peter, Marilyn's husband, was about to be transferred to Germany for business, they decided to purchase a lot and begin the process of finding a modular company to build their new home for their return to the United States. During the several years the Browns spent in Germany, Marilyn designed every aspect of the house. She knew she wanted a symmetrical Georgian and knew exactly how the house should be laid out. With the aid of a 3-D architectural design software, Marilyn laid out the entire house and even placed the furniture where she thought it would fit.

The Browns traveled extensively while in Europe and purchased some furnishings, such as a beautiful

Opposite page The rolling property at the rear of the house adds to the beauty of this structure.

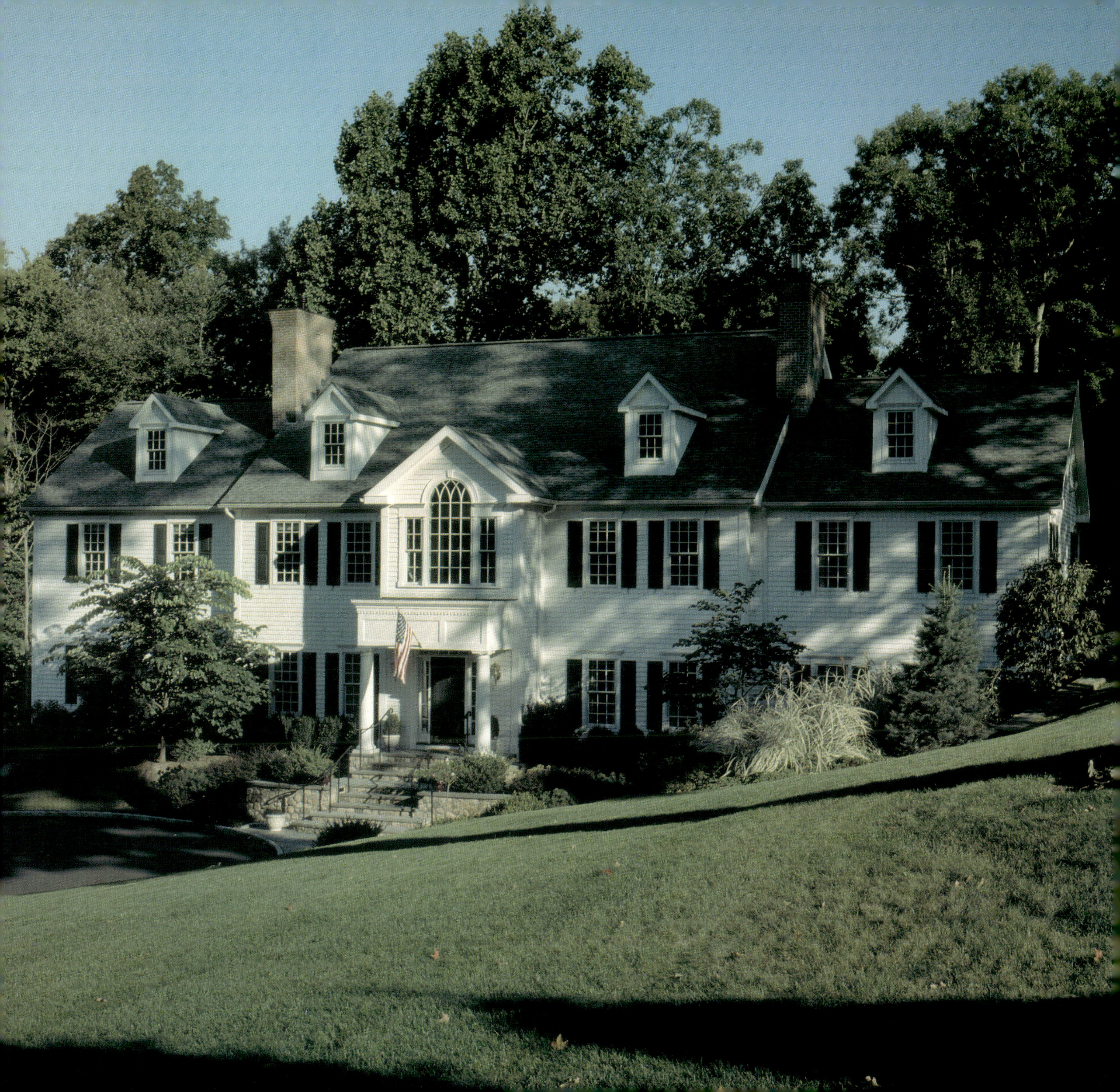

Below left Marilyn brought Persian rugs, such as this one in the living room, back from Germany.
Below right While visiting Versailles, Marilyn got the idea to cover the chandelier chain with decorative cord.

Manufacturer: *New Era Building Systems, Inc.*
Builder: *Winchester Modular Housing*
Interior designer: *Carol Andresen, Andresen Interiors*
Square feet: *7,000*
Number of boxes: *13*
Location: *New Canaan, CT*

grandfather clock from the Black Forest area to bring back for their new home. Marilyn was able to purchase Persian rugs at an excellent price because the very large rugs she wanted were typically too big for the more condensed German homes and not in demand in Germany.

When they returned to the United States, the builder they had been planning to employ was no longer able to accommodate their time schedule.

With the scale of the master bedroom so large, Marilyn opted for an oversized four-poster bed, which is also at a perfect height for viewing the built-in television over the mantel.

The arch and columns over the bath were completed on-site.

Although only one column was needed for structure, a second was added for symmetry.

Since he worked exclusively with the modular company they had selected, they had to find another modular company and builder to construct their home. Ultimately they were very happy with both new choices.

The Browns decided to purchase the modular boxes unfinished and complete all of the flooring, molding, kitchen and even one of the ceilings on-site. The ceiling in the hallway was delivered with just the joists, and a tray ceiling was later built by the contractor.

Typical of Georgian style, the Browns' house has a balanced design; steeply pitched roof; multiple dormers; palladian windows; a tall brick

This spacious family room is where the family spends a good deal of leisure time.

chimney; uniform, double-hung, multi-paned windows; and classical details.

Having replaced all of the molding in her old house, Marilyn had developed an excellent knowledge of the various types and opted for an extensive use of trim throughout the new house. She chose to include chair molding, base molding, box panels and ceiling medallions. Marilyn loves columns and used them in several areas of the house. When it became necessary to have a support structure in the large kitchen, she chose to use

This guest room on the main floor was designed for Marilyn's mother. To make her feel more at home, one of her dressers sits at the entrance.

two columns, which would add symmetry to the room, instead of the required one.

Marilyn put up lawn chairs and watched the house being set. She says neighbors were very nervous about what the house would look like when they saw the trucks arrive with the modular boxes. When they saw the results a week later, she began to receive many compliments from the naysayers.

Marilyn later worked with a local interior designer, selecting fabrics and window treatments that would complement the design of the house. She says she selected wood-burning fireplaces for the dining room and family room because "those are places we like to linger, enjoy the crackling of the wood and feel the warmth on a cold winter day." In several of the other rooms, such as the living room and bedroom, she chose gas-burning fireplaces for convenience.

The results of all the years of design work and focus on details is a gorgeous and comfortable Georgian period home.

Telluride

THE TOWN OF TELLURIDE is nestled in a box canyon just six blocks wide and twelve blocks long, surrounded by the thirteen-thousand-foot peaks of the San Juan Mountains. It is located in southwestern Colorado, at an elevation of 8,725 feet. Telluride has no traffic lights for forty-five miles and has a free gondola—the only public transportation gondola in the country, which connects the town to the mountain village with a thirteen-minute scenic ride. Bridal Veil Falls, at the east end of Telluride, is Colorado's tallest free-falling waterfall. With eighty-four trails, Telluride is rated by *Ski Magazine* as one of the top ten ski resorts in North America. It is not only a popular winter retreat but also a favorite summer destination with excellent hiking, luxurious lodging, a multitude of restaurants, many clothing boutiques and a championship golf course.

An ecologically sensitive area, the town operates the Galloping Goose bus, which runs on renewable, biodegradable, vegetable-oil-based fuel.

Opposite page A gondola connects the town to the mountain village, offering wonderful views and eliminating the need for a car.

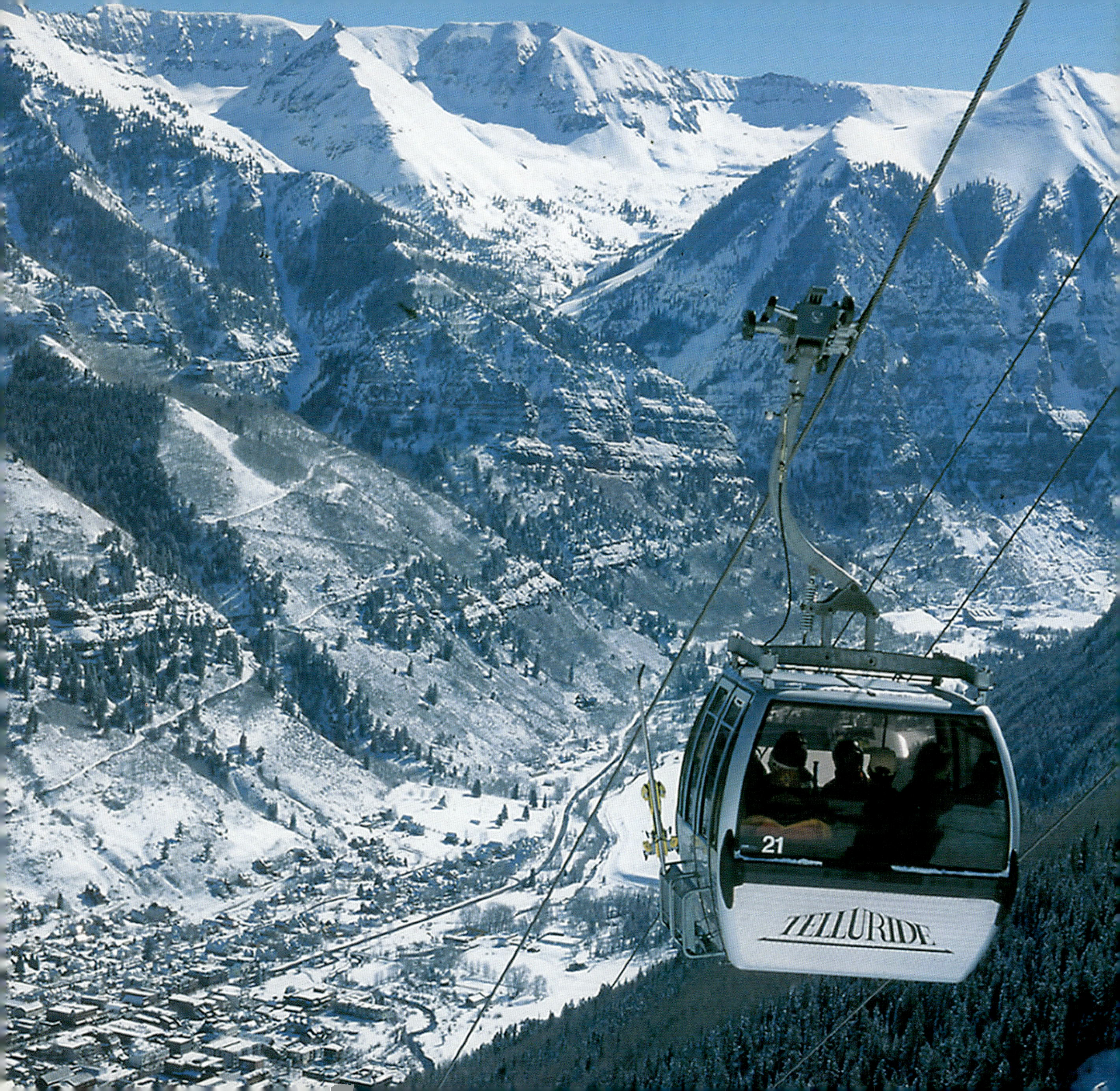
21
TELLURIDE

"no traffic lights
for forty-five miles and
a free gondola"

Above Elk stop by close to the Spiritsrest house to feast on lush grasses.
Right This tale of two cities shows the European modern look of the mountain village contrasted with the historic, Victorian mining town below.

With its long mining history, Telluride was designated as a National Landmark District in 1964. New buildings and remodel plans must be approved by the town's Historic and Architectural Review Board, which was set up to protect the Victorian and mining character of the town.

Telluride is also a mecca for festivals, such as the Telluride Film Festival, which has taken place for the past thirty Labor Day weekends, and the Telluride Blues & Brews Festival, which is held every September.

Left The fireplace has a gas burner because there is a ban on wood-burning fireplaces in Telluride. The expansive windows were used to bring in sunlight and the beautiful views.
Below The owners say food preparation is a participation sport in their home, so they created a space conducive to this activity.

Spiritsrest

Architect: *One Architects, Inc.*
Manufacturer: *Paragon Custom Homes*
Builder: *High Mark Development*
Square feet: *3,100*
Boxes: *9*

The owners of this mountain retreat owned a condominium in partnership with friends in Telluride for several years. When their friends moved to Europe, they decided to build their own home there because they enjoyed the area so much. The owners, who are in their mid-fifties and residents of Austin, Texas, spend

117

Below left The house was built on four levels with three short staircases.
Right All building materials are natural, with local Telluride Gold stone and cedar siding, as well as cedar shake roofing.

many winter vacations and the summer months in Telluride. In the winter they enjoy cross-country skiing, downhill skiing and snowshoeing, and in the summer they enjoy mountain biking and hiking. They take advantage of the cultural activities in town, including interesting scientific lectures, music festivals, film festivals and the friendly atmosphere of the small town.

The layout of the house was very important, and the owners worked diligently with their architects to incorporate all of their needs in the plan. They wanted to minimize the number of stairs, which they felt would make it more comfortable for them as they grow older. The hilly terrain dictated the need for several levels, resolved by creating several half levels with fewer steps between them. They also knew they would have guests and wanted the house to have private spaces for company as well as common areas where everyone could participate in food preparation, eating and spending time together.

They chose modular construction because they felt it was economically a better choice and it shortened the building time. The house was to be constructed during the winter months, which would have been more difficult with on-site construction. The owners also felt it was a better ecological choice because there would be less waste than would occur with traditional on-site stick construction.

This house was the third home the owners had built but their first modular construction. From their experience, they knew it was important to

Below This two-and-a-half-acre property is part of the Aldasoro Ranch, where building spaces were designated so neighbors wouldn't impact each other. This property was once part of a Basque sheep ranch. Today the development is in an elk calving area. Sheep still run on the adjacent ranch in the mountains above the development.

The owners felt it was an advantage to work with a local architect who understood local issues, would maximize the use of the lot and could deal with the nuances of the area, such as how the snow melts off the roof and the way the sun hits the house. The sun can be very intense during the day, even in the winter, so the correct placement of windows, for example, can take advantage of the sun's heat as well as the views.

With a modular home, most decisions must be made before the project begins, but for these owners this was an advantage since they could not be at the construction site regularly to make such decisions. This was one of the first houses that Paragon Custom Homes built that was not from their own plans. Because this was a new venture, they especially paid close attention to every detail of the construction, and the owners say the result was impeccable. They commented, "The most any marriage line [where the two boxes meet] was off was no more than one-eighth of an inch, which could easily be rectified."

specify all aspects of the house clearly so there would never be any questions about what they wanted. They prepared several binders with information on all electrical, appliance and finishing details. They included cut sheets and pictures with instructions on the locations for all outlets and fixtures. Copies of the binders were given to the architect, manufacturer and builder so that everyone was coordinated. The owners say their organization was praised by the builder, and the binders they created facilitated the whole process.

TELLURIDE

119

Below This beautiful bathroom includes a "double lady slipper" cast-iron bathtub, a 150-year-old antique trunk and white marble flooring.
Opposite top All of the tapestries adorning the walls were made by the owner.
Opposite bottom Glass blocks divide the master bedroom and the vanity behind it, which leads into a his-and-her bathroom with a steam shower between.

The owners credit the builder, Matt Mitchell, for ease of the construction process and the quality of the completed home. "Matt was optimistic, encouraging, always practical and wonderfully flexible working with clients." They say his reaction to their requests was always "Let's find a way to make that work."

The owners enjoy the elk and mule deer that graze on the surrounding land. They find it a very positive up-close and personal experience when they look out their window and see herds of elk wandering around only fifteen feet away.

Serapio Road House

MANUFACTURER: *Timber Creek Homes, Inc.*
BUILDER: *High Mark Development*
SQUARE FEET: *3,540*
NUMBER OF BOXES: *4*

This beautiful rustic house had to be built twice before the owner, Aleja, could move into it. When it was first built in October 2003, a gas leak developed in a weak pipe joint. There was a gas explosion, which demolished the house. This occurred just before Aleja was scheduled to move in. She says she decided to rebuild the house while it was still in flames. By October 2004, the exact duplicate of the first house was complete and she was finally able to move in.

The owner says she had previously built three other houses and knew exactly what she wanted, so she opted to design the house herself. She chose to build a modular house because of the cost savings and the fact that the house was mainly constructed in a controlled environment.

A spiral staircase leads from the master bedroom to a stained and stamped concrete deck, which is

Right The custom-designed door was made by local craftsman John Nelson. The exterior stone is local natural stone called Telluride Gold.
Opposite Page This beautiful, rustic-style home includes a Juliet balcony with metalwork designed by the owner. The railings on the second floor were found in a Colorado antiques store.

heated to alleviate snow buildup. Several of the rooms also have in-floor hydronic radiant heat to economize on heating bills and lessen the need for fossil fuels. At 9,750 feet above sea level, the temperature is consistently cold in the winter months, which is ideal for this type of heating system whose only flaw is its slow recovery time. Air-conditioning is not required in the warmer months at this high altitude; therefore, the venting usually needed for air-conditioning (and forced hot-air systems) is not necessary. The radiant heat offers a very clean, consistent type of heat and is the most commonly used system in the area, according to builder Matt Mitchell.

Building materials were generally all natural, from the green slate on the roof to the vintage barn wood from a winery in Napa Valley to the locally quarried natural stone siding. The "Juliet" balcony on the third level was built as a "drawer" and slid onto the house on-site. Aleja designed the custom metalwork on the balcony. She also choose all of the furnishings and incorporated several of her tapestries on the walls, adding a very warm and personal touch to this woodsy home.

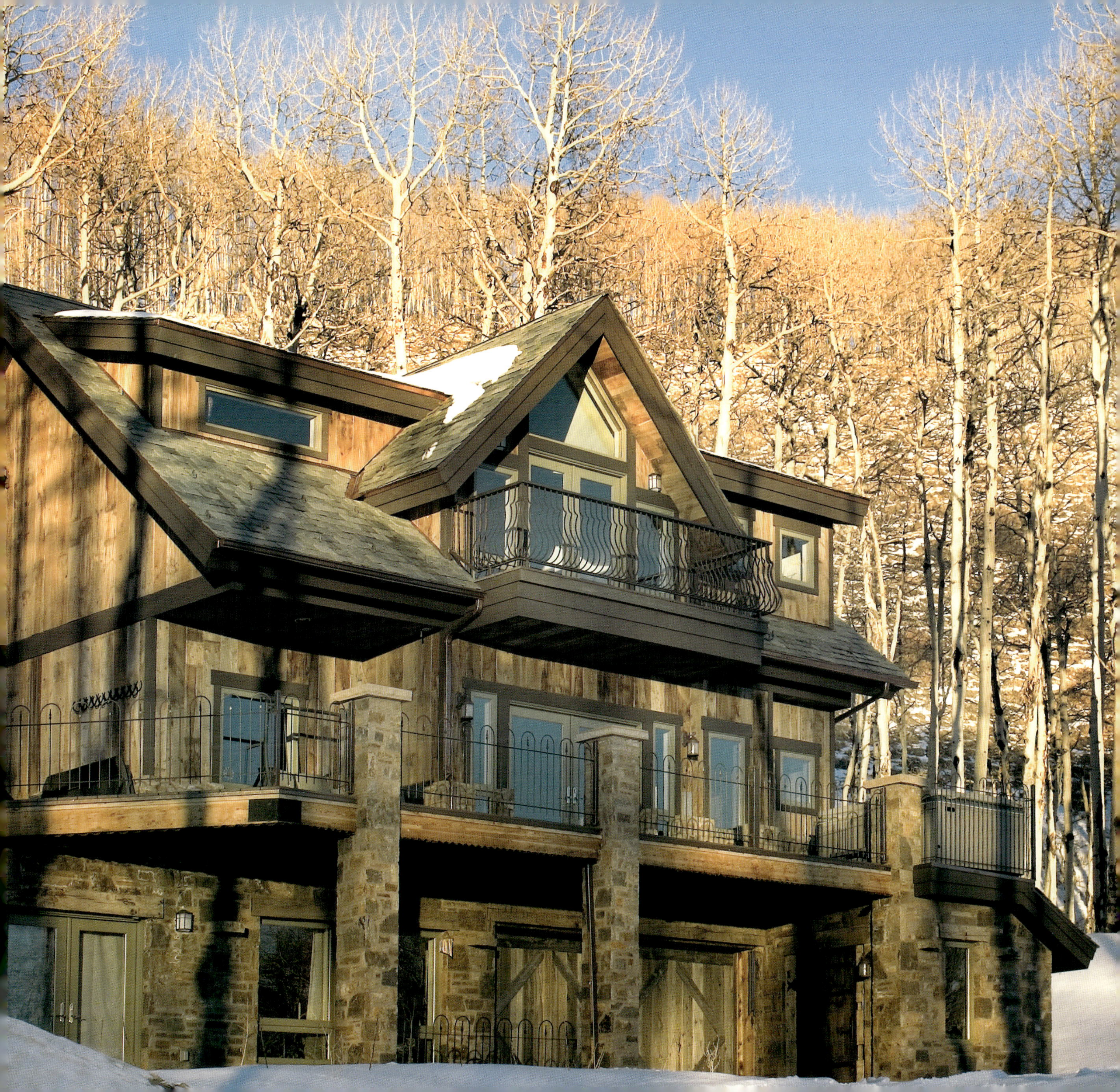

Juniper Hill House

MICHAEL GRUNBERG KNEW exactly what type of house he wanted and, after having built a modular home several years earlier, had an excellent understanding of the process. His prior experience taught him to do as much as possible in the factory to keep the costs down and to take full advantage of the modular concept. "Whatever it costs in the factory will be less than it costs to build on-site in the Northeast," says Michael. He wanted to build a very unique home with old-world charm and excellent materials that would last forever. He found a company that would cooperate and give him the customization he was looking for.

He opted to design the house himself, including the interior, and to hire his own contractors. Michael looked at every stone house in the area to decide how he wanted the stones to be laid. Ultimately he selected a native stone and chose a method of installing them. This gave the house its old-world look.

Opposite page The rear view of this house is enhanced by the beautiful landscaping and pool area.

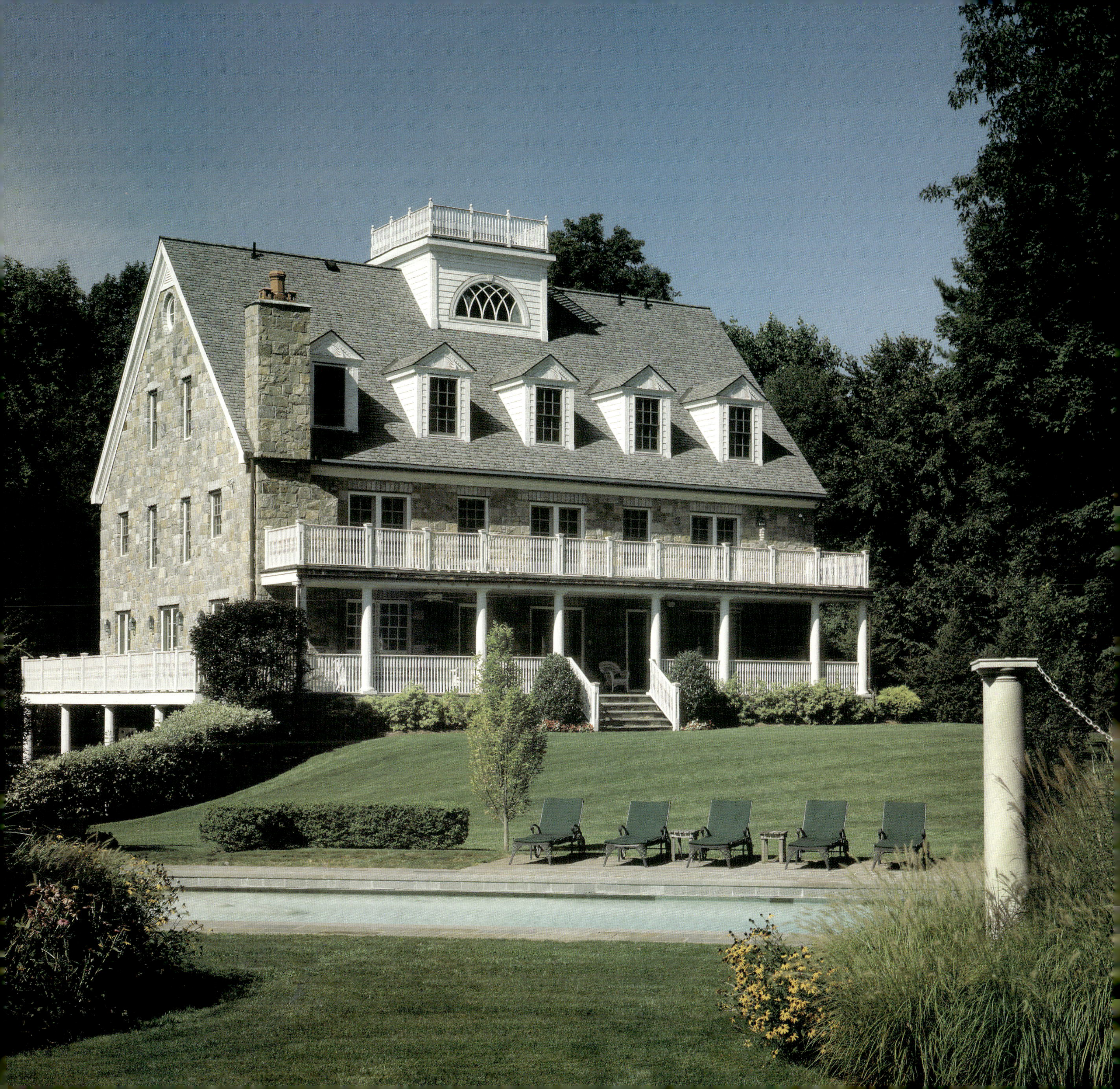

Right This stately front door is oversized and made of mahogany.
Below The exterior stones used on the house were hand-selected by the owners, as was the pattern in which they were laid.

Manufacturer: Huntington Homes, Inc.
Square feet: 9,000
Number of boxes: 10
Location: Greenwich, CT

The house was pre-wired in the factory for a security and audio system, as well as a backup generator. Because of the intricacy of this house, Michael says it would have taken two years to stick-build it instead of the one year it actually took.

The flooring has two layers of plywood that are overlapped, glued and screwed to give the house added strength. Michael looked for windows with true divided lights and wider mullions, like those used on the older homes he had seen. To protect the walls, he installed picture rails and chair rails throughout the house. All of the paintings from the Grunbergs' collection of nineteenth-century art hang from museum rods, which eliminate the creation of holes all over the walls.

The central hall has a lofty forty-five-foot ceiling, giving this house a tremendous feeling of space.

Below The formal living room is a great space to show off the nineteenth-century artwork.
Right The walls were upholstered to keep the room extremely quiet. The museum rods were covered with the same French-period-design fabric so they would blend with the walls. All of the furniture in the room is eighteenth century.

All of the doors in the house are oversized and solid wood, giving the house a grand appearance. Huntington Homes helped the Grunbergs by creating nonstandard door openings, both oversized and arched, in the factory. Because the roof of the house has such a high pitch (12/12), it was created as a separate section and set as a complete unit.

With so many boxes comprising

the house, it took two simultaneous weeks to set the house, as is the case with many larger modular constructions. The company sent the boxes in two separate trips from Vermont.

While the home was under construction, the Grunbergs, with their three children, lived in an old cottage on the property. One night, across the property, they saw a fire had erupted in their almost-completed new home. The floor scrapers had

To give the room an all-over wooden appearance, the cabinets were built in maple to match the wood of the floors.

sanded the floor, taking up a layer of polyurethane along with the sawdust. The polyurethane and dust ignited, creating a ten-foot hole in the master bedroom floor. Michael quickly put out the fire and the damage was minimal, with the smoke damage more extensive than the damage to the floor. Michael says that if it had been a stick-built house, it most likely would have burned down. There was less oxygen to fuel the flames because the house was so tight, and the Sheetrock used on the house was fire-rated. The space created between the two floors also prevented the flames from escaping to the floor below.

Michael, appreciating the attributes of the modular process, had planned

The arch in the master bathroom was created in the factory.

to put in a precast concrete modular foundation system, which is delivered cured, cutting down on the time required for it to set. Because the footings are attached to the wall system, it is also cost-efficient. Tampered stone, which is generally used with this system, was installed. The stone allows ground water to pass through and prevents the house from shifting. A traditional poured foundation system has separate footings and generally doesn't require the use of tampered stone. The Grunbergs later discovered that the company could not guarantee that the lip on the modular foundation system would hold the weight of the extensive rock facade. As a result, they ended up pouring a traditional foundation, with the tampered stone—ordinarily not used with a poured foundation—giving them an added level of stability. Michael believes this extra step helped prevent the house from shifting.

The result of this construction effort is a house that looks as if it has been there for a century and is extremely strong and magnificent to behold.

JUNIPER HILL HOUSE

The Colonial on Lake

ALTHOUGH THE ROSENS knew they wanted to build a custom home, they were not thinking modular. Gail Rosen was scanning magazines for kitchen ideas when she came upon an article about modular construction that caught her interest. Through word of mouth she heard about a beautiful modular home (the Juniper Hill House, page 124) that had been built in Greenwich, Connecticut. When she and her husband, Dan, went to explore this house, they became fast friends with the owners, who recommended they contact Huntington Homes. Dan was attracted to the savings in time and the accuracy with which modular homes are constructed. The Rosens visited the Huntington factory in East Montpelier, Vermont, and were impressed with the sophisticated equipment being used and the fact that the houses were built in a protected environment. Gail found that the designer at the company, Larry Roux, was able to interpret her cocktail-napkin sketch into the beautiful Colonial she had dreamed of.

Opposite page The large gazebo in the backyard has curtains all around that are spread out to block the sun at various times of the day.

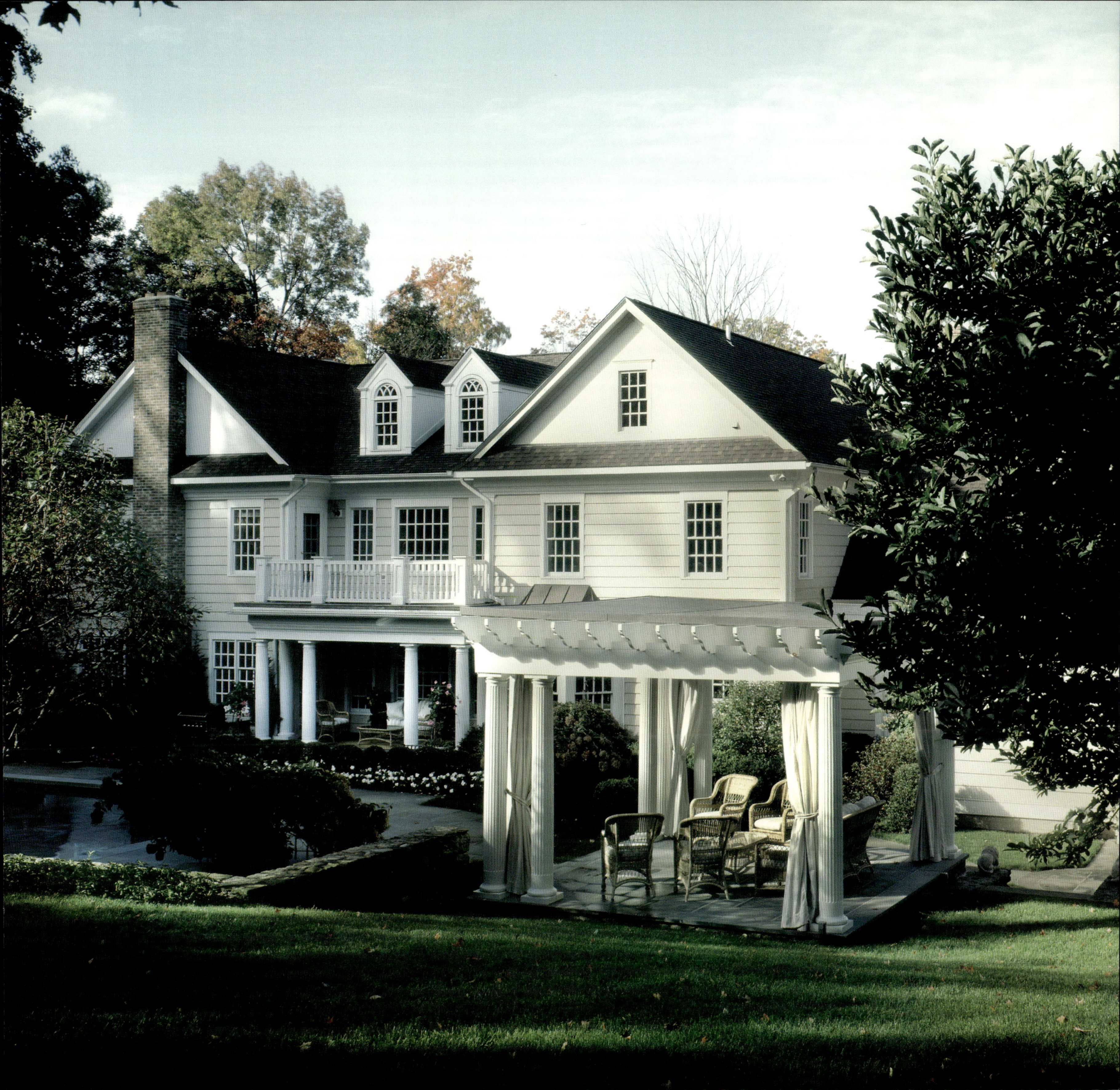

Below left The small stone lion on the coffee table is from a seventeenth-century French palace and is one of Gail's favorite antiques.

Below right The banquette is a cozy area for the family to gather and have meals in the evenings.

MANUFACTURER: *Huntington Homes, Inc.*
BUILDER: *Artisans, Inc.*
DESIGNER: *Larry Roux*
SQUARE FEET: *9,800*
NUMBER OF BOXES: *19*
LOCATION: *Greenwich, CT*

Very few changes needed to be made from the original design.

Of typical Colonial style, this house was designed with a symmetrical facade; paneled doors with sidelights; fanlight detailing; wood siding;

The double staircase creates an imposing look for the entrance of this stately Colonial. With the placement of the rear window, it is possible to see straight through the house to the beautiful pool fountain in the rear.

Below Passersby cannot believe this regal, beautiful Colonial is a modular home.
Right The tray ceiling in the master bathroom adds extra interest to this beautiful, spacious room.

multipaned, double-hung windows with dark green shutters; classical elements; and an elegant center entry hall.

After interviewing several contractors, the Rosens chose a builder who was a friend of the family and who typically built high-end on-site houses in Fairfield County. Although the contractor had never built a modular house before, the Rosens felt he had the expertise to deal with the specialty materials, such as the antique chestnut floors they were planning to use. The Rosens ordered the boxes empty without flooring, kitchen cabinets, bathrooms, etc., choosing to have most items completed on-site by the contractor. They knocked down a sixties Colonial at the site, which was demolished in about seven hours. Huntington had to make several trips to deliver all the boxes required to complete this house, but they were all set in a few days. Seven months later they were able to move into their home.

The garage in the house was built with double-high ceilings to accommodate the lift used to store Dan's favorite cars. The kitchen was built very spacious as a place for Gail to whip up her culinary delights and for

The marble on the island is all one piece—the largest slab the dealer had ever cut for a kitchen island.

family gatherings. Gail says, "We wanted to create a warm atmosphere in the kitchen, to feel almost like a den."

The house is enhanced throughout with Gail's etchings, paintings and photographs. Created over the years, they are unique and add a special touch to many of the rooms.

A darkroom was recently built in the basement as a place for Gail to develop her beautiful photographs.

The furnishings in the house are impeccable and a credit to Gail, who chose every item and fabric. What the Rosens created is a cozy, inviting home.

The pillars, door assembly, extensive molding and balcony make this entranceway grand.

Opposite page Multiple doors were used to open the inside space to the outside.

The Glide House

WHILE MICHELLE KAUFMAN was seeking affordable housing, her husband, Kevin Cullen, was determined that their home be eco-friendly. That meant minimizing the amount of energy used to build the house and the amount of energy necessary to maintain it, and to use renewable/recyclable materials. To solve both of their issues, Michelle designed the Glide House, a concept that can be adapted to meet other people's personal requirements in other locations as well.

Previously Michelle had worked with Frank Gehry, mainly designing museums. She has applied some of that experience to her residential designs by creating clean spaces and controlling sunlight at different times of the day. One of Michelle's favorite museums is the Norton Simon Museum in Pasadena, California. She loves the way the gallery spaces connect to the sculpture garden. Interested in creating that same synergy between the interior and exterior for her own home, Michelle

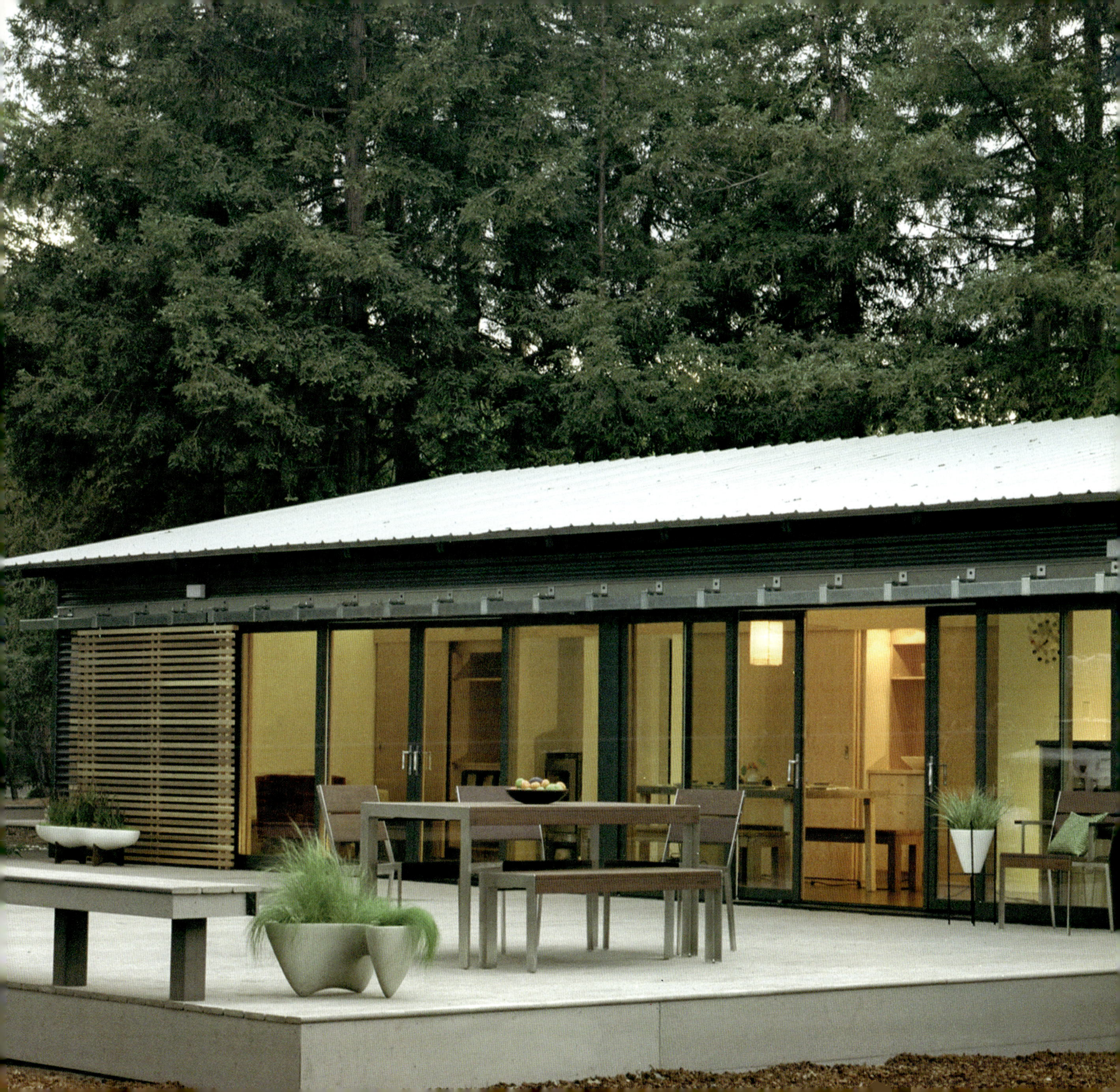

The placement of the dining table offers the best views of the outdoors and brings in the natural light, avoiding the unnecessary use of electricity.

> **Architect:** Michelle Kaufman, MK Architecture
> **Manufacturer:** Britco Structures, Ltd.
> **Project manager:** Construction Resource Group
> **Square feet:** 1,344
> **Number of boxes:** 2
> **Location:** Menlo Park, CA

incorporated this concept into her Glide House design.

"People are beginning to demand affordable hybrid or electric cars; affordable, healthy organic food; etc. The more people see that green construction is possible, the more the demand will increase, and the kinds of available choices in housing will increase," says Michelle. "I chose modular construction because it offers a product that is stronger, more energy efficient, yet more affordable than site-built construction. The costs are known up front (no surprises later), and it takes less time and a lot less hassle. It would be hard for me to ever go back to site-built construction now." She understood that there were design constraints with modular construction because of the limitations of dimension due to shipping, but she says, "Once someone understands those constraints and can design them in from the beginning, there aren't too many compromises to make in the design."

She chose materials for her house that required minimal maintenance, were renewable and created a

The freestanding gas stove is an energy-efficient method of supplementing the heating system in the house. A wood-burning fireplace is an alternative, although it's not permissible in some areas of California.

healthy environment. The exterior siding is corrugated Cor-Ten steel, which will rust over time and turn a deep red. The look requires no maintenance and has a warm velvet color and texture. "Sculptors, like my favorite, Richard Serra, have been using Cor-Ten steel for years. I enjoy watching people who can't help but touch these pieces," says Michelle. She also associates this material with home, having grown up in Iowa where there are a lot of "beautiful rusted steel buildings."

She designed the house with Low-E sliding doors that maximize cross ventilation, provide natural lighting and connect the interior of the house to the outdoors. Louvered panels in front of the glass doors allow the homeowners to customize the amount of daylight and shading, depending on the time of day and year. Operable clerestory windows opposite the doors maximize cross-ventilation breezes, and the indirect lighting created minimizes the need for electric lights. Depending on the location, this house can use solar panels, a wind generator or a hybrid system. In some cases, homeowners can hook up with their local utility to sell stored energy and then buy it back on cloudy days. "Through sustainable design and solar, geothermal or wind generator equipment, the Glide House can reduce, and sometimes eliminate,

THE GLIDE HOUSE

143

"they were determined that their home be eco-friendly"

The transom windows help create cross ventilation with the glass doors on the opposite side of the room.

utility bills," said Michelle. "Since the Glide House is not dependent on a local utility for all its power, it widens the range of potential building sites."

Galvalume standing-seam metal roofing was used because of its low maintenance and ability to blend in with solar panels, if they are used. A heat-recovery system recycles 30 percent of the energy used during heating and cooling. Radiant floor heating and high R-value insulation materials all contribute to the self-reliant nature of the system. Michelle worked with the Eagle Institute in Vancouver, Washington, to help incorporate healthy building technology—energy efficiency, healthy indoor air quality and an environment

The medicine cabinets were designed without frames to "disappear."

with the lowest possibility of mold. When complete, this house surpasses the Energy Star guidelines and meets or exceeds the guidelines of the American Lung Association Health House.

Strand-woven bamboo flooring was selected for its natural, textured beauty and durability. It is also a renewable resource that uses the waste from other bamboo flooring. Concrete countertops, made from recycled materials, create a beautiful and natural look and are also eco-friendly.

Wall space was designed on one side of the interior to create a flexible storage area for media, books, cooking objects, display and so on.

Friends of the architect were very interested in Michelle's design and she began to get inquiries from others who wanted similar designs. She then began to see the potential of mass-producing the Glide House with variations to suit personal requirements and needs in different areas of the United States and Canada.

She has designed two versions, one for snow conditions and one for non-snow conditions, like the one built in California. The difference is in the rooflines and windows, all designed to surpass insulation and seismic requirements. The basic box configurations and details remain the same in all of Michelle's designs, but depending on how the boxes are put together, she says, "you can have an L-shape, or a courtyard U-shape, or a long plan for a lot with views. There is quite a bit of flexibility, so the house can be configured to fit the site and the way the owner lives."

Victorian Nostalgia in New England

THIS BEAUTIFUL VICTORIAN home is in its greatest splendor in a bucolic autumn setting on a seven-and-a-half-acre parcel in Massachusetts. The Finns, with their two young boys, adorn the house in October with pumpkins, witches and everything Halloween, the family's favorite time of the year.

Steve Finn originally wanted a log home but decided against it because he felt it would not fit in well in the area of more traditional New England–style homes. He chose modular construction because he felt "it would be easier, less stressful and the time frame would be better." In November 1997, the Finns ordered their home; in May 1998, the house was set; and by the end of August of that year, they moved into their home, which took less than a year from start to finish to build.

The Finn house has many elements typical of the Queen Anne Victorian—steeply pitched gable, turret, varied roofline, tall chimney, large porch encircling

Opposite page The turret, wraparound porch, tall double-hung windows, varied rooflines and ornate brackets are all characteristic of Victorian style.

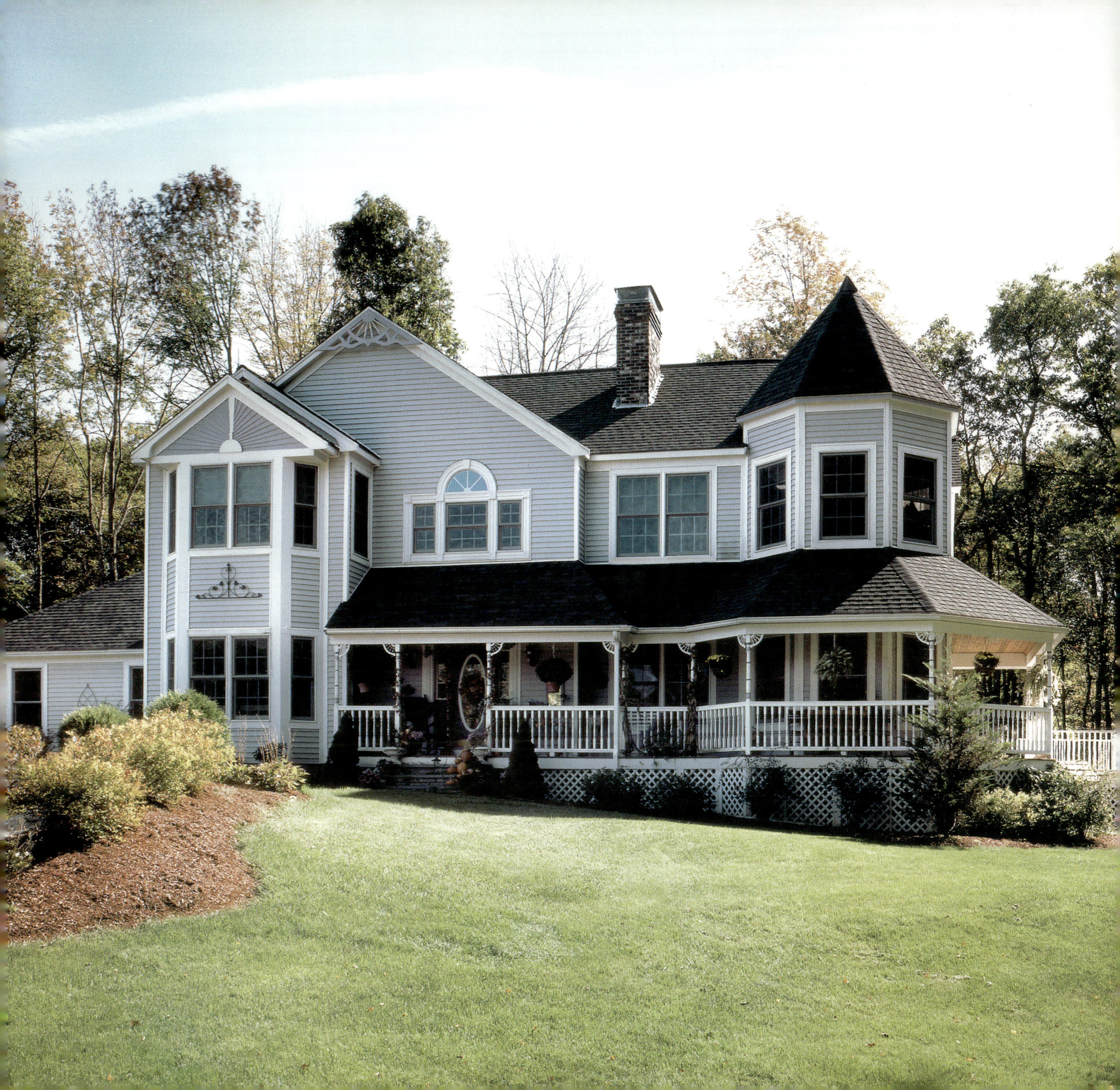

This inviting entrance has a double-high ceiling.

MANUFACTURER: *New England Homes, Inc.*
BUILDER: *Charles W. Boston*
DESIGNER: *Dave Wrocklage*
SQUARE FEET: *4,100*
NUMBER OF BOXES: *11*
LOCATION: *Boylston, MA*

much of the house, ornamental brackets and tall, double-hung windows.

Lori Finn chose a Victorian style from the many offered by New England Homes. She and her husband opted to purchase the home with some areas unfinished, since they had several craftsmen in the family and wanted to complete some of the work themselves. They worked with Charles Boston, their builder, to make alterations in the plans, site the house on the lot and help facilitate getting the house set. They opted to have the garage panelized, which is often done with modular homes, and the farm porch was stick-built on-site. Charles worked with the Finns to get the house started and continued as a consultant until the house was complete. However, Steve was familiar with construction and knew all the local tradesmen, having grown up in

The living-room fireplace, built on-site, is wood burning.

Because of the turret on the exterior of the house, the area over this romantic bed is domed.

"a wonderful combination of nostalgia and modern technology"

The warm burgundy color, lacy curtains, period chandelier and furnishings give the feeling of a more romantic time.

the area, so he worked as his own general contractor, supervising much of the construction. Steve and his two brothers, "gentlemen farmers," cleared the land, and then he contracted for the heating and ventilation systems and many of the other items necessary to complete the house.

Lori chose most of the furnishings, utilizing her memories of her grandparents' old Victorian house. To duplicate the look of their old tin ceiling, she chose textured wallpaper, which gives a similar appearance. She also did extensive research, reading many design magazines and books and religiously visiting home shows, picking up many style tips along the way. Lori says her home is "a work in progress" and that she will slowly continue to decorate as she gains insight and finds furnishings of interest.

The Finns' house is a wonderful combination of nostalgia and modern technology coming together to create a comfortable and attractive family home.

Hidden Pond Lane

THE OWNERS OF this Hampton shingle-type home live in Hong Kong with their four children. Both are lawyers—the father works for an American investment company, while the mother spends her time taking care of their four daughters. Each summer they return to Long Island to visit the grandparents and use the house only three months a year.

While the family is in Hong Kong, all four girls sleep in the same room because of the cramped living space in their Asian home; so despite the seven bedrooms they have in their Hampton house (including two in the basement), they all sleep in the same room here as well.

Whereas many of these gracious modular houses are custom built for a particular family, this house was built by a local builder as a spec house. Barry Altman, the principle of the building company for this project, believes in modular construction because the raw materials, working conditions and end product are superior. He says, "You get a little bit more house for a

Opposite page This farm-style cottage with its wraparound porch is typical of the casual architecture in the area.

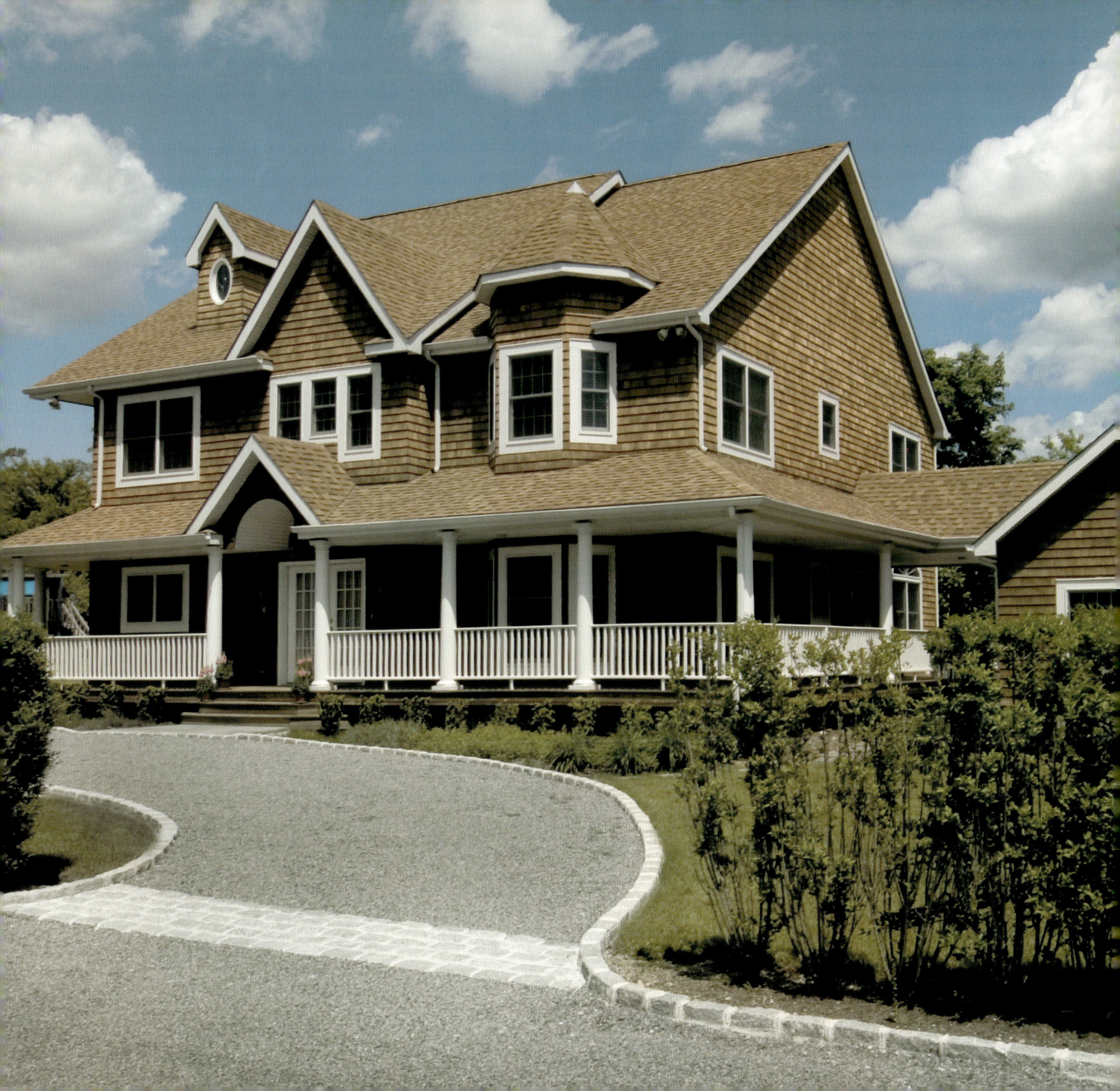

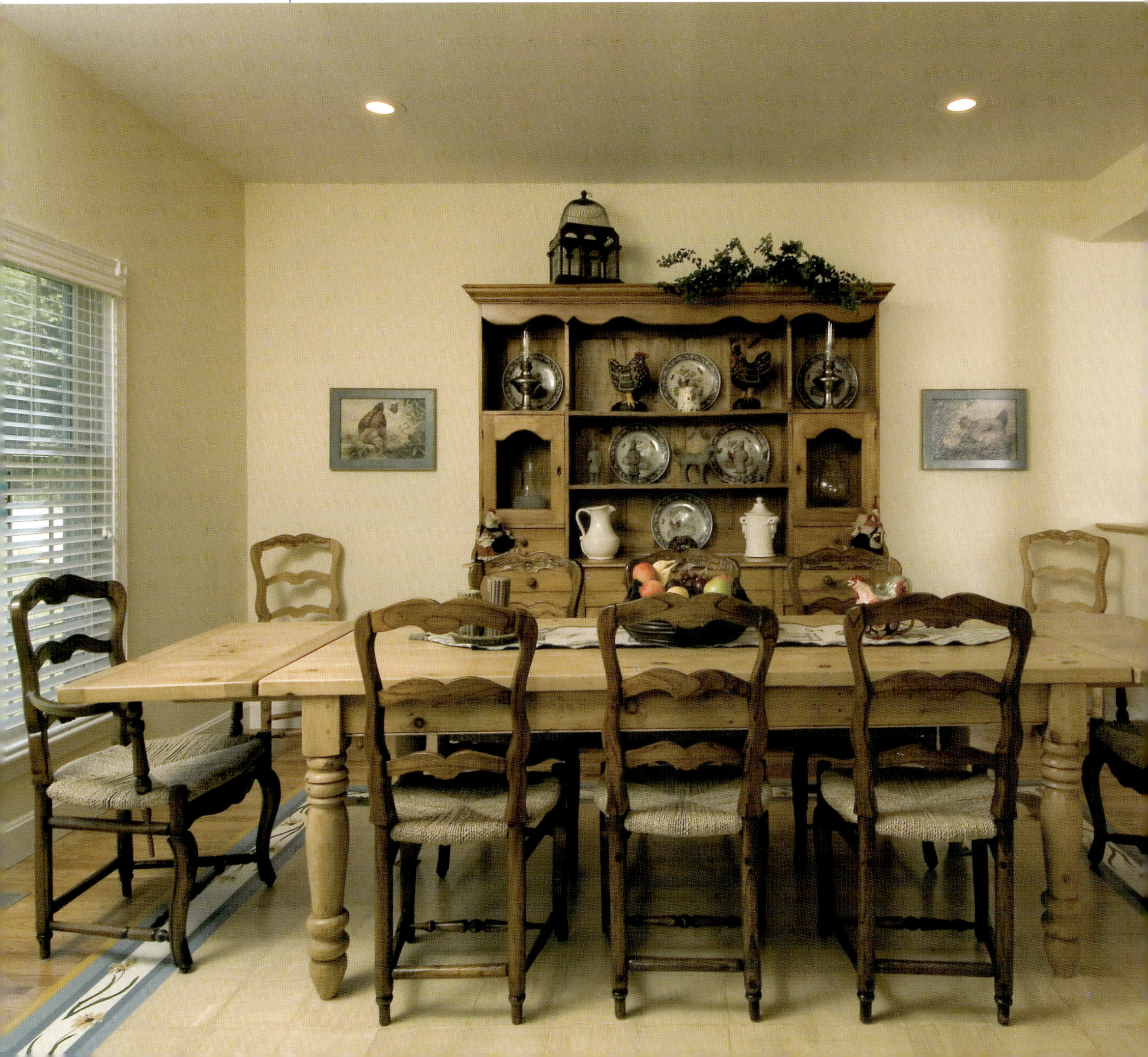

Opposite page The country dining area brings family and friends together.
Below The central staircase divides to separate the master suite from the other family bedrooms.

MANUFACTURER: *Sun Building Systems*
BUILDER: *Quality Crafted Homes*
INTERIOR DESIGNER: *Teri Seidman Interiors*
SQUARE FEET: *3,400*
NUMBER OF BOXES: *6*
LOCATION: *West Hampton, NY*

little bit less money in a shorter period of time." Although some stick builders, he says, may opt to use kiln-dried wood (the type most often used by modular companies), the advantageous properties of the wood are lost when the wood is exposed to harsh environmental conditions. Altman claims, "Mother nature is the enemy of every construction site." He believes the controlled environment in a modular factory creates more consistent working

HIDDEN POND LANE

155

Opposite page The open floor plan connects the various living spaces for an easy flow.
Right The french doors open to flood this living room with light.

conditions, allowing the builders to continuously work through hot, cold and wet weather, keeping workers in better spirits while they do—creating a better product.

Altman says that planning all aspects of the design may take a little more time because on-site decisions are more difficult to make with modular construction. With stick-built homes, homeowners may take a "wait and see" attitude, making decisions as they come about. It is more difficult to make decisions on modular homes once production has begun. On-site decisions are more difficult to make since most of the construction is completed in the factory prior to arriving at the site. This, however, avoids the costly change orders that often occur with site-built homes.

Although Altman concedes there are many excellent houses stick-built in the Hamptons and around the country, he believes he can offer a better, more consistent product to homeowners with modular construction. The sizes and styles of the houses have become larger and more sophisticated, but he believes the quality continues to be superior.

Bibliography

Arieff, Allison, and Bryan Burkhart. *Pre Fab*. Salt Lake City: Gibbs Smith, Publisher, 2002.

Buchanan, Michael. *Prefab Home*. Salt Lake City: Gibbs Smith, Publisher, 2004.

Carlson, Don O. *Automated Builder Dictionary/Encyclopedia of Industrialized Housing*, 3rd ed. *Automated Builder* magazine, 1995.

Gianino, Andrew. *The Modular Home*. North Adams, MA: Storey Publishing, LLC, 2005.

Resources

ARCHITECTS & DESIGNERS

Al Cappelli
792 Route 82
Hopewell Junction, NY 12533
845-226-7943

Bill Garnet
WJG & Associates
13999 Madison Avenue NE
Bainbridge Island, WA 98110
206-842-6171

Douglas Cutler, Architect
221 Danbury Road
Wilton, CT 06897
203-761-9561
www.modulararchitecture.com

Jon Brimus
30029 Last Dollar Road
Telluride, CO 81435
605-532-3444

Larry Link
L. J. Link Jr., Inc.
2201 Old Court Road
Baltimore, MD 21208
410-337-9528

Matt Mitchell
High Mark Development
120 Albert "J" Road
Telluride, CO 81435
970-728-4524

Michael MacDonald
49 Ridgewood Road
West Hartford, CT 06107
860-561-1542

Michelle Kaufman
MK Architecture
102-4th Street
Sausalito, CA 94965
415-999-4122
www.mkarchitecture.com

One Architects, Inc.
P.O. Box 3442
Telluride, CO 81435
970-728-8877
www.onearchitects.com

Porter Clapp Architects
180 Varick Street
New York, NY 10014
212-352-1210

Resolution: 4 Architecture
150 West 28th Street
Suite 1902
New York, NY 10001
212-675-9266
www.re4a.com

BUILDERS

All Homes
P.O. Box 7242
Newburgh, NY 12550

Artisan Homebuilders
143 Rowayton Avenue
Rowayton, CT 06853
423-510-1655
www.artisanshomebuilders.com

Bob Fisher General Contractor
7 Deer Run Road
Newburgh, NY 12550
914-542-1364

The Cedars of Town & Country
A Division of
Town & Country Cedar Homes
4772 US 131 South
Petoskey, MI 49770
800-968-3178
www.cedarhomes.com

Charles W. Boston
86 East Main Street
Westborough, MA 01581
508-366-1142
www.cwboston.com

Construction Resource Group
7960-170th Avenue NE
Redmond, WA 98052
877-550-2741
www.fastbuilding.com

Darrell Hoss Builders
41 Severance Drive
Stamford, CT 06905
203-316-0780

Edward A. Sandor, Builder, LLC
17 Fado Lane
Cos Cob, CT 06807
203-869-0352

The Fournier Group
208 DW Highway
Meredith, NH 03253
603-279-8900

Nantucket Island Homes
P.O. Box 192
Nantucket, MA 02554
508-228-4123

Quality Crafted Homes
(A Division of Custom Modular Homes of Long Island)
175 Montauk Highway
Water Mill, NY 11976
631-726-9300

Winchester Modular Housing
380 New Hartford Road
Barkhamsted, CT 06063
860-738-1782

Consultants & Organizations

Aleja of Telluride
aleja@telluridecolorado.net

Building Systems Network
www.buildingsystemsnetwork.com

Canada Homes Direct
www.net5000.com/canada/

Canadian Manufactured Housing Institute
150 Laurier Avenue West
Suite 500
Ottawa, ON, Canada K1P 5J4
613-563-3520
www.cmhi.ca

Eagle Institute
12300 SE 15 Street
Vancouver, WA 98683
877-882-9177
www.eagle-institute.com

Modular Building Systems Association
3029 North Front Street
Third floor
Harrisburg, PA 17110
717-238-9130
www.modularhousing.com

Modular Building Systems Council
National Association
of Home Builders
1201-15th Street, NW
Washington, DC 20005
800-368-5242
www.nahb.org

Modular Center
www.modularcenter.com

Interior Designers

Andersen Interiors
65 Nursery Road
New Canaan, CT 06840
203-972-2661

Bartlett's Home Interiors
6499 M 66 N
Charlevoix, MI 49720
231-547-2884

Joe Fanelli Unlimited
6 Charles Street, #5
New York, NY 10014
212-727-8546

L. J. Link, Jr., Inc.
2201 Old Court Road
Baltimore, MD 21208
410-337-9528

Teri Seidman Interiors
150 East 61st Street
New York, NY 10021
212-888-6551
www.teriseidman.com

Modular Manufacturers

Ameri-Log Homes
888-374-3025
www.ameri-loghomes.com

**Britco
Factory Built Buildings**
9267-194th Street
Surrey, BC, Canada V4N 4G1
800-527-4826
www.britco.com

Customized Structures, Inc.
272 River Road
P.O. Box 884
Claremont, NH 03743
800-523-2033
www.customizedstructures.com

Epoch Homes
Route 106
P.O. Box 235
Pembroke, NH 03275
877-463-7624
www.epochhomes.com

Haven Homes, Inc.
554 Eagle Valley Road
Beech Creek, PA 16822
570-962-2111
www.havenhomes.com

Huntington Homes, Inc.
344 Fassett Road
East Montpelier, VT 05651
802-479-3625
www.huntingtonhomesvt.com

New England Homes, Inc.
270 Ocean Road
Greenland, NH 03840
800-800-8831
www.newenglandhomes.net

New Era Building Systems
451 Southern Avenue
Strattanville, PA 16258
877-678-5581
www.new-era-homes.com

Paragon Custom Homes
600-12th Street SE
Madison, SD 57042
800-924-2926

Quality Engineered Homes, Ltd.
R.R. 2
Kenilworth, ON, Canada
N0G 2E0
866-276-5397
www.qualityhomes.ca

Sun Building Systems
#9 Stauffer Industrial Park
Taylor, PA 18517
570-562-0110
www.sunmodular.com

Timber Creek Homes
One Timber Creek Trail
Stratton, NE 69043
308-276-2478
www.timbercreekhomesinc.com

Westchester Modular Homes
30 Reagans Mill Road
Wingdale, NY 12594
800-832-3888
www.westchestermodular.com

Photographic Credits

All photographs are by Philip Jensen-Carter (pjcpixtkr@optonline.net) with the exception of the following:

Glen Graves
Pages 100–5

Kelly Jemison
Pages 116 (top left), 119 (right)

Michelle Kaufman
Pages 140–45

Alan Cuenca/Studio Cuenca
Pages 114 (center and bottom), 117–19, 120–23

Telluride Ski & Golf Company
Telluride Ski & Golf Company,
www.tellurideskiresort.com.
Photographers:
Doug Berry, page 114 (top)
Sven Brunso, page 115
Tony Demin, page 116 (center)

Verne McGrath
Pages 68–73

Publications

Automated Builder
805-642-9735
www.automatedbuilder.com

Building Systems
3352 South Newcombe Street
Lakewood, CO 80227
303-985-3564
www.buildingsystems.com